Study of biocompatible and biological materials
Can they be influenced by external factors?

by
Emilia Pecheva

The book gives an overview on biomineralization, biological, biocompatible and biomimetic materials. It reveals the use of biomaterials alone or in composites, how their performance can be improved by tailoring their surface properties by external factors and how standard surface modification techniques can be applied in the area of biomaterials to beneficially influence their growth on surfaces.

Important in studying of biomineralization is the study of the surface and chapter 2 explores typical techniques for surface characterization and shows how these techniques can be modified to serve specific needs in the study of biomaterials. Chapters 3 and 4 reveal factors that can be used to influence the growth of the biomaterial hydroxyapatite (the main inorganic constituent in mammal bones and teeth), namely laser energy, organic matrix and incorporation of minor amount of nanoparticles into the hydroxyapatite matrix. Proteins are also used to modulate the cellular interactions with the hydroxyapatite.

The following three chapters (5, 6 and 7) are devoted to an example of the pathological mineralization, namely the formation of bacterial films on teeth and soft tissues in the mouth and how they can be removed to achieve better oral health.

Study of Biocompatible and Biological Materials

Can they be Influenced by External Factors?

by

Emilia Pecheva

Biocompatible Materials group,

Institute of Solid State Physics, Bulgarian Academy of Sciences

72 Tzarigradsko Chaussee blvd., 1784 Sofia, Bulgaria

tel. +359 29795699, fax +359 2 4169357

emily@issp.bas.bg

Published by **Materials Research Forum LLC**
Millersville, PA 17551, USA

Published as part of the book series
Materials Research Foundations
Volume 11 (2017)
ISSN 2471-8890 (Print)
ISSN 2471-8904 (Online)

Print ISBN 978-1-945291-24-1
ePDF ISBN 978-1-945291-25-8

Distributed worldwide by

Materials Research Forum LLC
105 Springdale Lane
Millersville, PA 17551
USA
http://www.mrforum.com

Manufactured in the United State of America
10 9 8 7 6 5 4 3 2 1

Table of Contents

Preface

Living organisms have amazing ways to produce high-performance inorganic minerals of biological origin, such as bone, tooth enamel and dentin, cartilage, chitin, dental calculi, kidney/salivary/urinary stones, deer antlers, tendons of mammals, etc. The process by which these minerals with fine structure and high mechanical properties have been created in Nature is known as biomineralization. It involves a complex interaction between inorganic ions and organic molecules, resulting in a perfectly controlled nucleation and growth from aqueous solutions on specific active sites or microenvironments, which suppose stimulation of the crystal growth at some functional sites and inhibition or prevention on others. Stimulated by the fascinating natural materials, researchers are working in the labs to develop synthetic, biomimetic materials and composites based on the methods and systems existing in Nature. As a result, various biomaterials are created and used to replace parts of a living system or to function in intimate contact with living tissue. They are put in contact with living tissues and/or biological fluids to evaluate, treat, modify forms or replace a tissue, organ or function of the body. However, in spite of the tremendous achievements of modern science and technology, the Nature's ability to assemble inorganic compounds into astonishing hard tissues is still not achievable by the synthetic procedures.

Biomineralization is important in wide aspect of biological events, starting from the formation of bones and teeth, oyster shells, corals (physiological biomineralization) and expanding to pathological biomineralization, such as the diseases osteogenesis imperfecta, kidney stones, arteriosclerosis, myositis ossificans and etc. Current biomedical questions force researchers to focus their efforts on processes such as the occurrence, formation and degradation of calcium phosphate minerals in living organisms. The formation of calcium phosphate minerals in the Nature is obeyed to basic physico-chemical mechanisms for crystal formation, where conditions such as supersaturation, temperature and pH are important. Since the first biomaterials have been created, the scientists have been interested in investigating the surface properties of biomaterials, the interactions proteins-surface, as well as in modification of the surfaces to yield desired biological reactions. Tailoring the surface through a definite technique has been used for selective improvement of the biomaterials functions and effective lifetime by means of changing the physical, chemical, mechanical and other properties connected to the biocompatibility and functionality of the materials.

Chapter 1 of this book gives an overview on biomineralization, biological, biocompatible and biomimetic materials. It reveals the use of biomaterials alone or in composites, how their performance can be improved by tailoring their surface properties by external factors and how standard surface modification techniques can be applied in the area of biomaterials to beneficially influence their growth on surfaces. Important in studying of biomineralization is the study of the surface and chapter 2 explores typical techniques for surface characterization and shows how these techniques can be modified to serve specific needs in the study of biomaterials. Chapters 3 and 4 reveal factors that can be used to influence the growth of the biomaterial hydroxyapatite (the main inorganic constituent in mammal bones and teeth), namely laser energy, organic matrix and incorporation of minor amount of nanoparticles into the hydroxyapatite matrix. Proteins are also used to modulate the cellular interactions with the hydroxyapatite. The following three chapters (5, 6 and 7) are devoted to an example of the pathological mineralization, namely the formation of bacterial films on teeth and soft tissues in the mouth and how it can be removed to achieve better oral health. The development of *in vitro* bacterial models of dental plaque has been presented in chapter 5 and its disruption using the physical effect of cavitation occurring around dental instruments has been described in chapter 6. Attention to the efficacy of commercial toothpastes and toothbrushes on plaque removal has been paid in chapter 7.

Acknowledgements

In my work I rely very much on people and their professionalism. Mutual collaboration with numerous experts in the field of: biomaterials, materials science, laser science, chemistry, biology, microbiology, dentistry, medicine, etc. are the basis of my research and I enjoy it! Here in this book is the place to say a huge THANK YOU to these people; the list is long and I hope if I leave somebody out, he/she will forgive me:

My partner and colleague Assoc. Prof. Dr. T. Petrov from the Institute of Solid State Physics, Bulgarian Academy of Sciences who believes in me and supports me not only in my everyday life but also at work. He is the one that continually pushed me to write this recapitulation of my work.

Late Assoc. Prof. Dr. L. Pramatarova from the Institute of Solid State Physics, Bulgarian Academy of Sciences who believed in me when I was a PhD student and led me in research and life as if I was her daughter.

Prof. D. Nesheva, Dr. Z. Aneva and Assoc. Prof. Dr. N. Minkovski from the Institute of Solid State Physics, Bulgarian Academy of Sciences;

Prof. D. Dimova-Malinovska and M. Kamenova from the Central Laboratory of Solar Energy and New Energy Sources, Bulgarian Academy of Sciences;

Assoc. Prof. Dr. N. Krasteva and Dr. K. Hristova from the Institute of Biophysics and Biomedical Engineering, Bulgarian Academy of Sciences;

Dr. O. Sabotinov from Pulssvet Ltd., Sofia, Bulgaria;

I. Kostadinov and A. Penev from Pulslight Ltd, Sofia, Bulgaria;

Prof. G. Altankov from the Institut de Bioenginyeria de Catalunya;

Dr. P. Montgomery from the Instrumentation et Procédés Photoniques, Laboratoire ICube, Unistra-CNRS, Strasbourg, France;

Dr. P. Laquerriere from the INSERM, CNRS, Reims, France;

Dr. R. Sammons, Prof. D. Walmsley, Dr. W. Palin, Dr. R. Shelton and Dr. M. Hoffmann from the School of Dentistry, University of Birmingham, Birmingham, UK;

Prof. T. Hanawa and Prof. E. Kobayashi from the Institute of Biomaterials and Bioengineering, Tokyo Medical and Dental University, Tokyo, Japan;

Prof. A. Kolitsch, Prof. M. Helm, Dr. M. F. Maitz, Dr. M. T. Pham and Dr. A. Kondyurin from the Institute of Ion Beam Physics and Materials Research, Forschungszentrum Dresden-Rossendorf, Germany;

Prof. U. Bismayer from the University of Hamburg, Hamburg, Germany;

Dr. R. Presker and Prof. M. Stutzmann from the Walter Schottky Institute, Technical University of Munchen, Munchen, Germany;

Prof. R. Kniep and Dr. U. Schwarz from the Max Planck Institute for Chemical Physics of Solids, Dresden, Germany;

Dr. F. Riesz, Dr. A. L. Toth, Dr. E. Horvath and Prof. Csaba Balazsi from the Research Institute for Technical Physics and Materials Sciences, Hungarian Academy of Sciences, Budapest, Hungary;

Prof. I. Mihailescu and Dr. C. Lungu from the National Institute for Lasers, Plasma and Radiation Physics, Bucharest, Romania;

Dr. E. Palcevskis from the Semiconductor Physics Institute, Latvian Academy of Sciences, Riga, Latvia.

Prof. A. Iglic from the Faculty of Electrical Engineering, University of Ljubljana, Ljubljana, Slovenia.

Of equal importance are the funding organizations that supported my research and I would like to acknowledge their contribution to my research work:

the National Scientific Research Fund of the Bulgarian Ministry of Education (grants L1213/2002-2005; youth grant MUF1505 (2005-2006), TK-X-1708/2007-2010 and DFNI_Б01_18/2012-2015);

the Agency of Innovation of Bulgaria (grant NIF 02-54/2007-2010);

the Bulgarian Academy of Sciences (bilateral grants with the Academies of Sciences of Hungary, Romania, Poland, Latvia and Slovenia);

the European Foundation through grants MCF Training Sites MCF HPMT-CT-2000-00182 (2001-2002); Rhenaphotonics INTERREG III EU (2003-2006), Eureka EU project E!3033- Bionanocomposit (2006-2011), European Integrated Activity of Excellence and Networking for Nano and Micro-Electronics Analysis ANNA 026134-RI3 (2007-2010);

the North Atlantic Treaty Organization (NATO reintegration grant CBP.EAP.RIG.982693/2007-09);

the Centre national de la recherche scientifique CNRS-France (PICS 4848/2009-11);

the Japanese Society for the Promotion of Science (grant JSPS 18-06728/2006-2008);

the Engineering and Physical Sciences Research Council of the UK (EPSRC grant EP-J014060/2013-2015).

Abbreviations

AES	Auger electron spectroscopy
AFM	atomic force microscopy
ALP	alkaline phosphatase
α-MEM	alpha modified minimal essential medium
AMP	2-Amino-2-methyl-1-propanol
AMPSO	((1,1-Dimethyl-2-hydroxyethyl)amino)-2-hydroxypropanesulfonic acid sodium salt
ANOVA	univariate analysis of variance
AP	acid phosphatase
BSA	bovine serum albumin
CA	contact angle
CaP	calcium phosphate
CCD	charge coupled device
CG	cover glass
CNT	carbon nanotube
CPM	coherence probe microscopy
DI water	distilled water
DIC	differential interference contrast
DLC	diamond-like carbon
DMEM	Dulbeccos modified Eagle medium
DND	detonation nanodiamond
ECM	extracellular matrix
ED	electrodeposition
FBS	fetal bovine serum
FDA	fluorescein diacetate
EDS, EDX	energy dispersive X-ray spectroscopy
FFOCT	full field optical coherence tomography

FITC-FN	fluorescein isothiocyanate-labeled fibronectin
FN	fibronectin
FTIR	Fourier transform infrared spectroscopy
HA	hydroxyapatite
HDR	high dynamic range
HIS	high-speed imaging
HMDS	hexamethyldisiloxane
hs	horizontal soaking
HV	Vickers hardness
IR	infrared
LLSI	laser-liquid-solid interaction
LM	light microscopy
LSCM	laser scanning confocal microscope
LVDT	linear variable differential transformer
ND	nanodiamond
NIST	National Institute of Standards and Technology
OES	optical emission spectroscopy
PBS	phosphate buffered saline
PMMA	poly(methyl methacrylate)
PPHMDS	plasma polymerized hexamethyldisiloxane
QHPI	Quigley Hein plaque index
RBS	Rutherford backscattering
rms	root mean square
SBF	simulated body fluid
SCE	saturated calomel electrode
SCL	sonochemiluminescence
SEM	scanning electron microscopy
SG	silica glass

Si	silicon
SLV	scanning laser vibrometry
SNR	signal to noise ratio
SS	stainless steel
TB	toothbrush
TEM	transmission electron microscopy
Ti	titanium
TP	toothpaste
TSA	tryptone soya agar
TSB	tryptone soya broth
VN	vitronectin
vs	vertical soaking
WLSI	white light scanning interferometry
XPS	X-ray photoelectron spectroscopy
XRD	X-ray diffraction

CHAPTER 1

Introduction

Abstract

The field of biomaterials science has developed rapidly. It could be considered a multidisciplinary subject due to the involvement of disciplines such as physics, chemistry, medicine, dentistry, materials science, etc. It can also be considered interdisciplinary since these separate subjects come together and merge their boundaries within this area. Biomaterials science is one of those areas of advanced technology where a large number of terms have been introduced or invented. This chapter gives explanation of the most important and fundamental of these terms based on agreed definitions.

Keywords

Biomaterial, Biological Material, Biomimetic, Biomineralization, Composite, Biomaterial Surface Modification, Laser Modification of Biomaterials

Contents

1.   Biomaterials and Biological Materials

A number of definitions have been developed for the term "biomaterials". According to the experts in this field, biomaterials are "synthetic or natural materials used to replace parts of a living system or to function in intimate contact with living tissue" [1,2].

Biomaterials are put in contact with living tissues and/or biological fluids to evaluate, treat, modify or replace any tissue, organ or function of the body [3] and are now used in a number of different applications throughout the body [3,4]. Biomaterials are different from biological materials. Biomaterials are well accepted by living tissue and used for tissue replacement, while biological materials are produced by biological systems Such materials are for example: bones, tooth enamel, tooth dentin, cartilage, chitin, dental calculi, kidney/salivary/urinary stones, etc. In addition, there is the term biomimetic materials, which are synthetic materials but have composition, structure and properties similar to those of biological materials. The term biomimetics (or "the mimicry of life") was invented by the American engineer and biophysicist Otto Herbert Schmitt (1913 – 1998) in the 1950-s [5]. Biomimetic materials are created based on the methods and systems existing in nature and used to study, design and construct new engineering systems, materials, chemical compounds and modern technology. In spite of the tremendous achievements of modern science and technology, nature's ability to assemble inorganic compounds into astonishing hard tissues such as shells, teeth, bones, antlers, skeletons, spicules, etc. is still not achievable by modern synthetic procedures. This is not surprising – designs found in nature are the result of millions of years of evolution and competition for survival.

Living organisms have amazing ways to produce various high-performance materials and over 60 different inorganic minerals of biological origin have already been found [2]. Among them, calcium phosphates (CaPs) are of a special importance since they are the most important inorganic constituents of hard tissues in vertebrates [6,7]. In the form of a non-stoichiometric, ion-substituted and calcium deficient hydroxyapatite (HA, referred to as "biological apatite"), CaPs are present in bones, teeth, deer antlers and tendons of mammals to give these organs stability, hardness and function [6,8,9]. Current biomedical questions of persistent pathological and physiological mineralization in the body force researchers to focus their efforts on processes such as the occurrence, formation and degradation of CaPs in living organisms [10-12].

Natural bone is the most typical calcified tissue of mammals. It is a nanocomposite material consisting of mineral, water, collagen, noncollagenous proteins, lipids, vascular elements and cells, whose absolute amounts vary with animal age, tissue site, health and dietary status. Bone serves two essential functions: on one hand as a structural material which is able to support its own weight, withstand acute forces, bend without shattering, etc., and on the other hand as an ion reservoir for both cations and anions. Both functions depend to a significant extent on the exact size, shape, chemical composition and crystal structure of the mineral crystallites, and their arrangement within an organic matrix. The major inorganic component of bone mineral is a biological apatite, which might be defined as a poorly crystalline, non-stoichiometric and ion substituted calcium-deficient HA (chemical formula $Ca_{10}(PO_4)_6(OH)_2$). From the material point of view, bone can be considered as an assembly of distinct levels of hierarchical structural units from macro-through micro- and to nanoscale to meet numerous functions [2-10]. Over a lifetime, bone is continuously remodeling itself: old bone is eroded away and new bone is deposited. Mechanically, the ratio of matrix to mineral concentration is very important. Newly deposited bone is low in mineral density. As mineral density increases and bone crystals increase in size, the stiffness of the bone increases. Bone crystallinity is also a measure of turnover rate, where the average crystal size is smaller in young bone because newly formed crystals are smaller than mature crystals.

Table 1. Composition of enamel and dentin in teeth (volume % of total tissue)

Component	Enamel	Dentin
CO_3-HA	85	47
Water	12	20
Protein and lipid	3	33

Teeth are another CaP-based calcified tissue of vertebrates, which consist of two biological materials: enamel (the crown which is the top part of the tooth) and dentin (bellow the crown and above the tooth root) [13]. The tooth main component is mineral carbonated HA (CO_3-HA); it also contains water and organic components, such as proteins and lipids (Table 1) [14]. Enamel is built mainly of highly crystalline CO_3-HA (85 vol%) and has less water and organics, which makes it much harder than the dentin. On the other side, dentin has more water and organics as constituents, and only 47 vol% CO_3-HA, which makes it more fragile. In dentin, the nanodimensional building crystals (~25 nm width, ~4 nm thickness and ~35 nm length) of biological apatite are smaller than those of enamel. Dentin is analogous to bone in many aspects; for example, it has a similar composition and a hierarchical structure up to the level of bone lamellae [13].

2. Composites and Advantages

"Composite materials" (or composites for short) are engineered materials made from two or more constituent materials with different physical or chemical properties and which remain separate and distinct on a macroscopic level within the complete structure [15]. There are two major categories of constituent materials in composites: a matrix (or a continuous phase) and a dispersed phase. The continuous phase is responsible for filling the volume, as well as it surrounds and supports the dispersed material by maintaining its relative position. The dispersed phase is usually responsible for enhancing one or more properties of the matrix. Most of the composites target an enhancement of mechanical properties of the matrix, such as stiffness and strength; however, other properties, such as erosion stability, transport properties (electrical or thermal), radio-opacity, density or biocompatibility, might also be of great interest. The synergism between the major and minor components produces properties, which are unavailable from the individual constituent materials [16]. Moreover, by controlling the volume fractions, local and global arrangement of the dispersed phase, the properties and design of the composites can be varied and tailored to suit necessary conditions.

Biomimetic HA is highly biocompatible inorganic material from the apatite family and integrates well with bones due to its chemical resemblance to mammalian bone and teeth [14]. However, the benefits of HA coatings are constrained by the low strength of adhesion to the underlying surface and by the limited cohesion within their layers. It is considered that the addition of a second (minor) phase might improve the mechanical strength and adhesion of the apatite coatings [17]. General information on the major fabrication and processing techniques in the area of biomaterials can be found elsewhere [18].

3.   External Factors for Tailoring the Surface Properties of Biomaterials

Stimulated by fascinating natural examples, such as bones, teeth, cartilage, shells and corals, researchers are working in the labs to develop synthetic, biomimetic nanocomposites by simulating the basic principles of biological mineralization (biomineralization). Following this goal, many groups have been exploring the potential of a simple immersion method in an aqueous supersaturated solution, known as simulated body fluid (SBF) [19] in order to mimic the process of biological apatite formation. A disadvantage of the method is the long time required to produce CaP coatings. Therefore, a new trend of applying external energy to stimulate the HA formation *in-vitro* has evolved: ultrasound and electrical fields, ultraviolet and microwave irradiation have been used for this purpose [20–22]. Fang et al. [20] applied ultrasound to accelerate the formation of HA in aqueous solutions at low temperature and managed to reduce the time for the HA formation from hours to minutes. Another research group [21] reported the formation of bio-resembling apatite crystals from SBF on TiO_2 under illumination with ultraviolet light and following immersion of the samples for 10 days in the SBF. There is also a tendency of using microwave energy for acceleration of specific chemical reactions [21]; under the microwave influence, precipitation of HA from aqueous medium was obtained within an hour. In 2004, a novel method involving laser–liquid–solid interaction (LLSI) for the facilitated synthesis of HA by using a laser irradiation of materials simultaneously immersed in the supersaturated SBF has been proposed for the first time [23-25] and adopted by other research groups [26]. This process provides a plasma state generated by the transient temperature increase induced by optical absorption of the laser light by the matter placed in the liquid and is a combination of a physical process induced by the laser and a subsequent chemical process in the liquid. For laser irradiation onto a substrate placed in an aqueous solution supersaturated with respect to HA, such a transient process induces the deposition of HA nuclei onto the laser-irradiated area. More attention on the LLSI process will be given in Chapter 3.

4. Surface Modification Techniques of Biomaterials

The first biomaterials were created at the end of 1940 and the beginning of 1950. Since this time, the scientists have been interested in investigating the surface properties of biomaterials, the interactions proteins-surface, as well as the modification of the surfaces. Thus, the idea of how the surface can be tailored to yield desired biological reactions has arisen [27].

Important in the study of biomineralization is the investigation of the surface and its appropriate structural and/or chemical modification leading to desired bioreactions and to improving the biomaterial performance, since they are subjected to different (aggressive) body fluids. Tailoring the surface through a definite technique is used for selective improvement of the biomaterial functions and effective lifetime by means of changing the physical, chemical, mechanical and other properties connected to the biocompatibility and functionality of the materials [28]. The surface properties are completely different and more complex than these of the bulk material, and since most of the biological reactions are taking place at the surface, it determines in a large extend the biological behavior of the materials.

The methods used for surface modification with the aim to coat the surface with a biocompatible HA layer and to improve the biocompatibility of the materials can be divided according to their nature to physical, chemical and electro-chemical methods [29]. Physical methods include vacuum plasma spraying of HA on the surface, ion sputtering, ion implantation, pulsed laser deposition, dipping of the material in a CaP precursor solution (a biomimetic approach), and many others [30-46]. Among the electro-chemical methods are anodic oxidation, electro-crystallization, electrodeposition, electrophoretic deposition, etc. [39-41]. The chemical methods such as surface-induced mineralization, immersion in NaOH followed by heat treatment, preliminary formation of CaP nuclei on the material through immersion in Ca-containing solution, acid etching, oxidation with H_2O_2, sol-gel techniques, alternate dipping process, etc. are the closest to the actual biological process [42-44].

Physical methods for coating the surface with HA and based on treatment with energetic ion beams, such as ion implantation, ion-beam deposition and plasma spraying have proved themselves very successful since they offer favorable modification of the surface properties without affecting the bulk of the material, and as an addition they improve the adhesion of the surface layer. These techniques combine flexibility and low temperatures of the process with excellent control, reliability and reproducibility. Particularly ion implantation finds extensive application in the biomaterials area for creating functional surfaces with tailored biological activity [28]. This technique utilizes a beam of highly energetic ions to modify the structure of the surface (layer with thickness about 100 nm) and its chemical composition at low temperature [45]. It does not change the bulk properties of the materials mainly due to the low temperatures at which they are kept. The process can be applied on any material – metal, ceramic, polymer, etc., but its effects are specific for the particular material due to differences in the structure. Micropatterning or the creation of design on the surfaces is a widely used method for controlling the microenvironment and the structure of the crystal growth sites, for increasing the surface area, thus ensuring a possibility for better connection with the bone [46]. It can be performed by ion and molecular beam etching, allowing patterning on a nanoscale, however, these techniques are extremely slow and complicated.

In recent years, increased interest in exploiting the possibilities of the simple and cheap method of immersing the materials in a supersaturated CaP aqueous solution (SBF) resembling the ion composition, concentrations and pH of the human blood plasma at low temperatures ($< 100^0$C) is observed [19,30-32]. This method is maximizing the approach to the real conditions in nature at which minerals like CaPs are formed [19,47]. One of the approaches in this method is the mechanical or chemical modification of the material surface before its immersion in the SBF. Thus, the opportunity for precipitation of homogeneous, dense and having good adhesion CaP layers is higher. Other advantages are the possibility to obtain biomimetic coating on materials with complex shapes and hence better implant connection with the bone, as well as the influence of high temperatures on the substrate required in some techniques is avoided. Although the increased number of studies, the exact mechanism, by which coatings are produced with this method, is not yet clear enough. The process of precipitation is complicated due to the possibility for a formation of a few solid phases depending on the solution composition and its pH.

Modification of titanium (Ti) by chemical treatment with NaOH and subsequent thermal treatment yields activation of the Ti surface by spontaneous formation of bioactive surface apatite layer through which the metal implant is connected and integrated in the bone [48]. The formation of such layer is ascribed to the sodium titanate ($Na_2Ti_5O_{11}$)

formed on the surface of the metal as a result of the NaOH and thermal treatment, and which hydrolyzes at immersion of the metal in SBF to form Ti hydrogel ($HTiO_3$-nH_2O). The assumed mechanism is based on the process of ion exchange between the metal surface and the solution: Na^+ ions are released from the metal surface to the solution and H_3O^+ ions go from the solution to the surface. As a result, Ti-OH groups are formed on the surface which charge it negatively (at pH around 7.4) and attract Ca^{2+} ions from the solution, thus forming calcium titanate ($Ca_3Ti_2O_7$) which is positively charged. In turn, Ca^{2+} ions attract PO_4^{3-} ions from SBF forming apatite nuclei on the surface. Once formed, they keep growing spontaneously by consuming more Ca^{2+} and PO_4^{3-} from the already saturated with respect to apatite SBF. Investigations with solutions containing only calcium or phosphate ions show that the deposition of Ca^{2+} ions is prerequisite for the formation of apatite nuclei on Ti [49,50].

5.   Laser Modification of Biomaterials

The idea of using laser light has been based on the fact that over the past decades, laser sources of energy have proven to be major tools for surface processing of a large variety of metals, ceramics and polymers, as well as for processing of living tissues and synthesis of biocompatible materials. Lasers have been utilized as exceptional contactless, fully adaptable sources of clean energy. Applications in research, medicine and dentistry, such as pulsed laser deposition and processing of HA [35-38], biofilms modification [51], bone defect healing [52], modification of root canal dentine [53], laser sterilization of dental implants [54] are already well established. Lasers have also been employed to increase the long-time performance of Ti dental implants through assuring cleanliness, specific micro-relief and a stable oxide layer on the surfaces [55]. Lasers have been successfully used since the mid-1960s for eye surgery through the precise photocoagulation of the retina [56]. Particularly, pulsed lasers have unique qualities that can be applied to process biomaterial thin films. Pulsed lasers have proven to be an invaluable tool in the research and development of new thin film materials because of the unique nature of the laser-material interaction. Laser processing utilizes the high power density available from focused and localized laser sources to melt, heat, or modify the material on and near the surface [57]. Depending upon the particular material system and process parameters, it may involve only modification of microstructure, grain refinement, phase transformations, alloying and mixing of multiple materials, and mixing and formation of composite system on the surface without actually affecting the bulk material itself. Laser processing does not require a specific environment: it can be conducted in air at extremely rapid speed with high precision and repeatability [57]. An area of interest is the laser surface modification (or texturing) of biomaterial surfaces to give them desired features. Laser micromachining to achieve 3D structures on micrometer and submicrometer scales is rapidly proving to be a good alternative to standard process such as photolithography [58]. Conventional lithographic techniques involve the need of photoresists and thin film resists. These are time-consuming steps and the use of resist development and other solvents alter the polymer surface texture, chemistry, etc. On the contrary, laser-based techniques do not require organic solvents that modify the structure of the polymer-treated surface or denature proteins present on the surface, neither require clean or dark rooms, spin coaters, photoresist, etc. When the substrate material is difficult to be removed or when the surface geometry is complex, lasers seem to be the ideal solution. Lasers have the ability to generate extremely complex microstructures/features with high resolution: if a laser beam is coupled with a noncontact mask and optical systems, it is possible to produce submicrometer features. Micromachining and laser microtexturing have been carried out to produce relatively uniform microgrooves on a range of materials [58,59]. Laser ablation can be used for various surfaces [60,61]. The use of lasers for texturing surfaces presents many advantages: they are rapid and extremely clean sources of energy, they are useful for the selective modification of surfaces.

References

[1]   D.F. Williams: The Williams dictionary of biomaterials (Liverpool University Press, Liverpool, UK, 1999).

[2]   S. Mann: Biomineralization: Principles and Concepts in Bioinorganic Materials Chemistry (Oxford University Press, New York, USA, 2001).

[3]   International standard definition ISO/EN, Consensus Conference, Chester, UK, 1992.

[4]   K.D. Jandt, Evolutions, revolutions and trends in biomaterials science – a perspective, Adv. Eng. Mater. 9 (2007) 1035-1050. https://doi.org/10.1002/adem.200700284

[5]   J.M. Benyus: Biomimicry: innovation inspired by nature (William Morrow, New York, USA, 1997).

[6]   H.A. Lowenstam, S. Weiner, On Biomineralization (Oxford University Press, New York, USA, 1989).

[7]   M. Vallet-Regí, J.M. González-Calbet, Calcium phosphates as substitution of bone tissues, Prog. Solid State Chem. 32 (2004) 1–31. https://doi.org/10.1016/j.progsolidstchem.2004.07.001

[8]   S. Weiner, L. Addadi, Design strategies in mineralized biological materials, J. Mater. Chem. 7 (1997) 689–702. https://doi.org/10.1039/a604512j

[9]   S. Weiner, H.D. Wagner, The material bone: structure-mechanical function relations, Ann. Rev. Mater. Sci. 28 (1998) 271–298. https://doi.org/10.1146/annurev.matsci.28.1.271

[10]   J.D. Pasteris, B. Wopenka, E. Valsami–Jones, Bone and tooth mineralization: why apatite? Elements 4 (2008) 97–104. https://doi.org/10.2113/GSELEMENTS.4.2.97

[11]   C.M. Giachelli, Ectopic calcification: Gathering hard facts about soft tissue mineralization, Am. J. Pathol. 154 (1999) 671–675. https://doi.org/10.1016/S0002-9440(10)65313-8

[12]   T. Kirsch, Determinants of pathological mineralization: crystal deposition diseases, Curr. Opin. Rheumatol. 18 (2006) 174–180. https://doi.org/10.1097/01.bor.0000209431.59226.46

[13]   S.J. Nelson, Wheeler's Dental Anatomy, Physiology and Occlusion, in: W.B. Saunders (Ed.), Philadelphia, USA, 2009, p. 368.

[14]   R. LeGeros, Calcium phosphates in enamel, dentine and bone, in: H.M. Myers (Ed.), Calcium phosphates in oral biology and medicine, Basel, Karger, 1991, vol. 15.

[15]   Information on https://en.wikipedia.org/wiki/Composite_material (assessed November 2016)

[16]   F.L. Matthews, R.D. Rawlings, Composite materials: engineering and science (CRC Press LLC, Boca Raton, 2000, p 480).

[17]   Sh. Catledge, M. Fries, Y.K. Vohra, Nanostructured surface modifications for biomedical implants, in: H. Nalwa (Ed.), Encyclopedia of Nanoscience and Nanotechnology, American Scientific Publishers, 2004, v. 10, pp. 1-22.

[18]   S.L. Evans, P.J. Gregson, Composite technology in load-bearing orthopaedic implants, Biomaterials 19 (1998) 1329-1342 https://doi.org/10.1016/S0142-9612(97)00217-2

[19]   T. Kokubo, H. Kushitani, S. Sakka, T. Kitsugi, T. Yamamuro, Solutions able to reproduce in vivo surface-structure changes in bioactive glass-ceramic A-W, J. Biomed. Mater. Res. 24 (1990) 721–734. https://doi.org/10.1002/jbm.820240607

[20]   Y. Fang, D. Agrawal, D. Roy, R. Roy, P. Brown, Ultrasonically accelerated synthesis of hydroxyapatite, J. Mater. Res. 7 (1992) 2294–2298. https://doi.org/10.1557/JMR.1992.2294

[21]   T. Kasuga, H. Kondo, M. Nogami, Apatite formation on TiO2 in simulated body fluid, J. Cryst. Growth 235 (2002) 235–240. https://doi.org/10.1016/S0022-0248(01)01782-1

[22]   E. Lerner, S. Sarig, R. Azoury, Enhanced maturation of hydroxyapatite from aqueous solutions using microwave irradiation, J. Mater. Sci. Mater. Med. 2 (1991) 138–141. https://doi.org/10.1007/BF00692971

[23]   L. Pramatarova, E. Pecheva, T. Petrov, N. Minkovski, A. Kondyurin, R. Pramatarova, Ion beam modified surfaces as substrates for hydroxyapatite growth induced by laser-liquid-solid interaction, Proc. SPIE 5449 (2004) 41–45. https://doi.org/10.1117/12.563091

[24]   L. Pramatarova, E. Pecheva, R. Presker, Formation of surfaces organized on both micro- and nanometer scale by a laser-liquid-solid-interaction process, Plasma Processes and Polymers 3 (2006) 248-252. https://doi.org/10.1002/ppap.200500135

[25]   E. Pecheva, T. Petrov, C. Lungu, P. Montgomery, L. Pramatarova, Stimulated in vitro bone-like apatite formation by a novel laser processing technique, Chemical Engineering Journal 137 (2008) 144-153. https://doi.org/10.1016/j.cej.2007.07.096

[26]   B-H. Lee, A. Oyane, H. Tsurushima, Y. Shimizu, T. Sasaki, N. Koshizaki, A New Approach for Hydroxyapatite Coating on Polymeric Materials Using Laser-Induced Precursor Formation and Subsequent Aging, Applied Materials and Interfaces 1 (2009) 1520-1524. https://doi.org/10.1021/am900183e

[27]   J. Breme, V. Biehl, A. Hoffmann, Tailor-made composites on titanium for medical devices, Adv. Engin. Mater. 2 (2000) 270-275. https://doi.org/10.1002/(SICI)1527-2648(200005)2:5<270::AID-ADEM270>3.0.CO;2-9

[28]   P. Sioshansi, E. Tobin, Surface treatment of biomaterials by ion beam process, Surf. Coat. Technol. 83 (1996) 175-182. https://doi.org/10.1016/0257-8972(95)02838-2

[29]   B. Feng, J. Chen, S. Qi, L. He, J. Zhao, X. Zhang, Carbonate apatite coating on titanium induced rapidly by precalcification, Biomater. 23 (2002) 173-179. https://doi.org/10.1016/S0142-9612(01)00093-X

[30]   F. Barrere, P. Layrolle, C. van Blitterswijk, K. de Groot, Biomimetic coatings on titanium: A crystal growth study of octacalcium phosphate, J Mater. Sci. Mater. Med. 12 (2001) 529–534. https://doi.org/10.1023/A:1011271713758

[31]   A. Oyane, M. Uchida, Y. Yokoyama, C. Choong, J. Triffitt, A. Ito, Simple surface modification of poly(e-caprolactone) to induce its apatite-forming ability, J. Biomed. Mater. Res. A 75A (2005) 138–145. https://doi.org/10.1002/jbm.a.30397

[32]   Y. Yokoyama, A. Oyane, A. Ito, Preparation of a bonelike apatite-polymer fiber composite using a simple biomimetic process, J. Biomed. Mater. Res. B App. Biomater. 86B (2008) 341–352.

[33]   K. de Groot, R. Geesink, CPAT Klein, P. Serekian, Plasma sprayed coatings of hydroxyapatite, J. Biomed. Mater. Res. 21 (1987) 1375–1381. https://doi.org/10.1002/jbm.820211203

[34]   Y.Z. Yang, K.H. Kim, J.L. Ong, Review on calcium phosphate coatings produced using a sputtering process - An alternative to plasma spraying, Biomaterials 26 (2005) 327–337. https://doi.org/10.1016/j.biomaterials.2004.02.029

[35]   C.M. Cotell, Pulsed laser deposition and processing of biocompatible hydroxylapatite thin films, Appl. Surf. Sci. 69 (1993) 140–148. https://doi.org/10.1016/0169-4332(93)90495-W

[36]   R.K. Singh, F. Qian, V. Nagabushnam, R. Damodran, B.M. Moudgil, Excimer laser deposition of hydroxyapatite thin films, Biomaterials 15 (1994) 522–528. https://doi.org/10.1016/0142-9612(94)90018-3

[37]   R. Narayan, Ch. Jin, A. Doraiswamy, I. Mihailescu, M. Jelinek, A. Ovsianikov, B. Chichkov, D.B. Chrisey, Laser Processing of Advanced Bioceramics, Adv. Eng. Mater. 7 (2005) 1082-1098. https://doi.org/10.1002/adem.200500155

[38]   F. Sima, C. Ristoscu, L. Duta, O. Gallet, K. Anselme, I.N. Mihailescu, Laser thin films deposition and characterization for biomedical applications, in: Rui Vilar (Ed.), Laser Surface Modification of Biomaterials, Woodhead Publishing, Elsevier, 2016, ch. 3, pp 77-125. https://doi.org/10.1016/B978-0-08-100883-6.00003-4

[39]   S. Yamaguchi, T. Yabutsuka, M. Hibino, T. Yao, Generation of hydroxyapatite patterns by electrophoretic deposition, J. Mater. Sci. Mater. Med. 19 (2008) 1419-1424. https://doi.org/10.1007/s10856-006-0053-6

[40]   X. Pang, I. Zhitomirsky, Electrodeposition of hydroxyapatite–silver–chitosan nanocomposite coatings, Surf. Coat. Technol. 202 (2008) 3815–3821. https://doi.org/10.1016/j.surfcoat.2008.01.022

[41]   P. Peng, S. Kumar, N. H. Voelcker, E. Szili, R. Smart, H. Griesser, Thin calcium phosphate coatings on titanium by electrochemical deposition in modified simulated body fluid, J. Biomed. Mater. Res. 76A (2006) 347–355. https://doi.org/10.1002/jbm.a.30514

[42]   T. Taguchi, A. Kishida, M. Akashi, Hydroxyapatite formation on/in hydrogels using a novel alternate soaking process, Chem. Lett. 8 (1998)711–712. https://doi.org/10.1246/cl.1998.711

[43]   P. Li, C. Ohtsuki, T. Kokubo, K. Nakanishi, N. Soga, K. de Groot, The role of hydrated silica, titania and alumina in inducing apatite on implants, Journal of Biomedical Materials Research 28, 1994, 7-15 https://doi.org/10.1002/jbm.820280103

[44]   K. Nishio, M. Neo, H. Akiyama, S. Nishiguchi, H-M. Kim, T. Kokubo, T. Nakamura, The effect of alkali- and heat-treated titanium and apatite-formed titanium on osteoblastic differentiation of bone marrow cells, J. Biomed. Mater. Res. 52 (2000) 652-661. https://doi.org/10.1002/1097-4636(20001215)52:4<652::AID-JBM9>3.0.CO;2-W

[45]   P. Townsend, J. Kelly, N. Hartley, Ion Implantation, sputtering and their applications (Academic Press, 1976).

[46]   N. Ozawa, T. Yao, Micropattern formation of apatite by combination of a biomimetic process and transcription of resist pattern, J. Biomed. Mater. Res. 62 (2002)579–586. https://doi.org/10.1002/jbm.10281

[47]   A. Campbell, G. Fryxell, J. Linehan, G. Graff, Surfaceinduced mineralization: a new method for producing calcium phosphate coatings, J. Biomed. Mater. Res. 32 (1996) 111-118. https://doi.org/10.1002/(SICI)1097-4636(199609)32:1<111::AID-JBM13>3.0.CO;2-P

[48]   B. Yang, J. Weng, X. Li, X. Zhang, The order of calcium and phosphate ion deposition on chemically treated Ti surfaces soaked in aqueous solution, J. Biomed. Mater. Res. A 47 (1999) 213-219. https://doi.org/10.1002/(SICI)1097-4636(199911)47:2<213::AID-JBM11>3.0.CO;2-C

[49]   Q. Feng, F. Cui F, H. Wang, T. Kim, J. Kim, Influence of solution conditions on deposition of calcium phosphate on titanium by NaOH-treatment, J. Cryst. Growth 210 (2000) 735-740. https://doi.org/10.1016/S0022-0248(99)00502-3

[50]   H. Wen, J. Wijn, Q. Liu, K. de Groot, F. Cui, A simple method to prepare calcium phosphate coatings on Ti6Al4V, J. Mater. Sci. Mater. Med. 8 (1997) 765-770. https://doi.org/10.1023/A:1018512612695

[51]   A. Richter, P. Gonprot, R. Smith, Biofilms and their modification by laser irradiation, Methods Phys. Res. B 180 (2001) 1–11. https://doi.org/10.1016/s0168-583x(01)00389-5

[52]   G. Guzzardella, M. Fini, P. Torricelli, G. Giavaresi, R. Giardino, Laser Stimulation on Bone Defect Healing: An In Vitro Study, Lasers Med. Sci. 17 (2002) 216–220. https://doi.org/10.1007/s101030200031

[53]   A. Ebihara, B. Majaron, L.-H. Liaw, T. Krasieva, P. Wilder-Smith, Er:YAG laser modification of root canal dentine: influence of pulse duration, repetitive irradiation and water spray, Lasers Med. Sci. 17 (2002) 198–207. https://doi.org/10.1007/s101030200029

[54]   M. Kreisler, W. Kohnen, C. Marinello, H. Gotz, H. Duschner, B. Jansen, B. D'Hoedt, Bactericidal effect of the Er:YAG laser on dental implant surfaces: an in vitro study, J. Periodont. 73 (2002) 1292–1298. https://doi.org/10.1902/jop.2002.73.11.1292

[55]   A. Karacs, A. Joob Fancsaly, T. Divinyi, G. Peto, G. Kovach, Morphological and animal study of titanium dental implant surface induced by blasting and high intensity pulsed Nd-glass laser, Mater. Sci. Eng. C 23 (2003) 431–435. https://doi.org/10.1016/S0928-4931(02)00316-8

[56]   F. L'Esperance Jr, An opthalmic argon laser photocoagulation system: design, construction, and laboratory investigations, Trans. Amer. Ophthalmol. Soc. Ann. Meeting 66 (1968) 827-904.

[57]   A. Kurella, A. Dahotre, Review paper: surface modification for bioimplants: the role of laser surface engineering, J. Biomater. Appl. 20 (2005) 5-50. https://doi.org/10.1177/0885328205052974

[58]   H. Reimers, J. Gold, B. Kasemo, D. Chakarov, Topographical and Surface Chemical Characterization of Nanosecond Pulsed-Laser Micro- machining of Titanium at 532 nm Wavelength, Appl. Phys. A 77 (2003) 491–498. https://doi.org/10.1007/s00339-002-1477-6

[59]   Y. Nakayama, T. Matsuda, Surface Microarchitectural Design in Biomedical Applications: Preparation of Microporous Polymer Surfaces by an Excimer Laser Ablation Technique, J. Biomed. Mater. Res. 29 (1995) 1295–1301. https://doi.org/10.1002/jbm.820291017

[60]   A.C. Duncan, F. Weisbuch, F. Rouais, S. Lazare, C. Baquey, Laser Microfabricated Model Surfaces for Controlled Cell Growth, Biosens. Bioelectr. 17 (2002) 413–426. https://doi.org/10.1016/S0956-5663(01)00281-0

[61]   A. Pique, R. Auyeung, J. Stepnowski, D. Weir, C. Arnold, R. McGill, D. Chrisey, Laser Processing of Polymer Thin Films for Chemical Sensor Applications, Surf. Coat. Techn. 163– 164 (2003) 293–299. https://doi.org/10.1016/S0257-8972(02)00606-0

CHAPTER 2

Characterization of Biomaterials and Improved Techniques for Better Imaging

Abstract

Nowadays there are numerous requirements for materials created in laboratories. In order to get a complete picture of the materials under study (i.e. are these materials appropriate for the particular application, have they been modified in a desired way, etc.) researchers need to obtain complementary information using various techniques. As there is no ideal material for any application, there is also no ideal technique. Any of the physico-chemical characterization techniques known today has its pros and cons. Some standard techniques are constantly under development by researchers in order to modify them to fulfill particular needs when they do not give optimal results. This chapter gives a brief overviews of standard imaging, structural and chemical characterization techniques related to the theme discussed, as well as reveals efforts being made to modify the white light interferometry technique for obtaining more informative imaging.

Keywords

Biomaterial Surface Characterization, Imaging Techniques, Structural and Chemical Analysis, Coherence Probe Microscopy Improvement

Contents

1. Physico-Chemical Characterization Techniques for Biomaterials

Light microscopy (LM) in materials analysis generally refers to reflected light microscopy [1,2]. In this method, light is directed vertically through the microscope objective and reflected back through the objective to an eyepiece, view screen, or camera. Transmitted light is occasionally used for transparent and translucent materials. For some low-magnification work, external, oblique illumination can be reflected off the sample into the objective. Magnification of the sample image is obtained by light refraction through a combination of objective lenses and eyepieces. The minimum feature resolution is approximately 0.2 µm. However, smaller features - as small as about 0.05 µm - can be detected by image contrast enhancement with polarized light, interference contrast, and dark field illuminations. The resulting images can be recorded either on traditional films or as digital files for computer display, analysis, and storage.

Bright field mode of the LM produces true color images at magnifications up to ~2000x. The sample surface is uniformly illuminated by incident light rays directed perpendicularly to the sample surface. Light reflected back toward the objective lens is collected and focused on the eyepieces to form the observed image. Surfaces that are reflective and perpendicular to the light rays appear bright. Alternatively, non-reflective or oblique features reflect less light and appear darker.

Polarized light mode produces enhanced contrast for features that have anisotropic refractive properties. Two polarizing lenses are inserted into the optical path - one in the incoming illumination and one between the sample and eyepieces. When these lenses are rotated 90° to one another, the "crossed polarizers" result in the subtraction of a portion of the light spectrum by destructive interference. Contrast is obtained between sample features that have different reflective properties. Many metals, including beryllium, zirconium and titanium, are anisotropic and exhibit grain contrast with polarized light illumination. Polarized light enhances contrast for many polymer samples and shows variations of internal stress in some clear polymers.

Differential interference contrast after Nomarski (DIC) mode produces a 3D image by creating brightness contrast on very minor topographical changes. DIC utilizes crossed polarizers as described for polarized light. A double quartz prism is also inserted into the light path to split the incident light into two separate paths. This results in two slightly shifted images of the sample on the viewing plane, which produces contrast between features with different heights and topographic orientations. The analyzer can be adjusted to obtain various degrees of interference to enhance selected features or create contrast colors in the image.

Darkfield mode gives enhanced contrast from subtle topographic features. An occluding disk is placed in the light path, blocking the direct vertical illumination. Peripheral rays in the illumination are reflected in such a way that light reaches the sample at oblique angles. The absence of incident vertical rays results in bright reflectance only from oblique features, such as ridges, pits, scratches and particles. Thus, subtle features that might be completely invisible in bright field microscopy are readily observed with this method.

Quantification is possible with LM because the microscope magnification is calibrated against reference standards. Lateral feature dimensions can be measured to accuracy greater than 0.5 µm. Computer analysis of digitally acquired images can measure area and volume fractions, particle sizes, grain size, and other features.

Typical applications of LM include small sample inspection, metal microstructure evaluation, small feature measurements, fracture mode identification, corrosion failure inspection and surface contamination evaluation. Sample size, shape, and condition requirements depend on the configuration of the microscope but typically large samples (cm x cm) can be measured without special sample preparation.

Atomic force microscopy (AFM) is a form of scanning probe microscopy where a small probe is scanned across the sample to obtain information about the sample's surface [1,2]. The information gathered from the probe's interaction with the surface can be as simple as physical topography or as diverse as measurements of the material's physical, magnetic, or chemical properties. These data are collected as the probe is scanned in a raster pattern across the sample to form a map of the measured property relative to the XY position. Thus, the AFM microscopic image shows the variation in the measured property, e.g. height or magnetic domains, over the area being imaged. The AFM probe has a very sharp tip, often less than 100 Å diameter, at the end of a small cantilever beam. The probe is attached to a piezoelectric scanner tube, which scans the probe across a selected area of the sample surface. Interatomic forces between the probe tip and the sample surface cause the cantilever to deflect as the sample's surface topography (or other properties) changes. A laser light reflected from the back of the cantilever measures the deflection of the cantilever. This information is fed back to a computer, which generates a map of topography and/or other properties of interest. Areas as large as 100 µm squares to less than 100 nm squares can be imaged this way.

In contact mode AFM the probe is scanned at a constant force towards the sample surface to obtain a 3D topographical map. When the probe cantilever is deflected by topographical changes, the scanner adjusts the probe position to restore the original cantilever deflection. The scanner position information is used to create a topographical image. Lateral resolution of <1 nm and height resolution of <1 Å can be obtained.

In tapping mode AFM the probe cantilever is oscillated at or near its resonant frequency. The oscillating probe tip is then scanned at a height where it barely touches or "taps" the sample surface. The system monitors the probe position and vibrational amplitude to obtain topographical and other property information. Accurate topographical information can be obtained even for very fragile surfaces. Optimum resolution is about 50 Å lateral and <1 Å height. Images for phase detection mode, magnetic domains, and local electric fields are also obtained with this mode.

In lateral force microscopy mode, the lateral deflection of the probe cantilever as the tip is scanned across the sample in contact mode is measured. Changes in lateral deflection represent relative frictional forces between the probe tip and the sample surface.

Phase detection microscopy: with the system operating in tapping mode, the cantilever oscillation is damped by interaction with the sample surface. The phase lag between the drive signal and actual cantilever oscillation is monitored. Changes in the phase lag indicate variations in the surface properties, such as viscoelasticity or mechanical properties. A phase image, typically collected simultaneously with a topographical image, maps the local changes in material's physical or mechanical properties.

In magnetic force microscopy mode, local variations in the magnetic forces at the sample's surface are imaged. The probe tip is coated with a thin film of ferromagnetic material that will react to the magnetic domains on the sample surface. The magnetic forces between the tip and the sample are measured by monitoring the cantilever deflection while the probe is scanning at a constant height above the surface. A map of the forces shows the sample's natural or applied magnetic domain structure.

Image analysis with AFM: since the images are collected in a digital format, a wide variety of image manipulations are available for AFM data. Quantitative topographical information, such as lateral spacing, step height, and surface roughness are readily obtained. Images can be presented as two-dimensional or three-dimensional representations in hard copy or as digital image files for electronic transfer and publication. Nanoindentation can be also performed with the AFM microscope as a specialized probe tip is forced into the sample surface to obtain a measure of the material's mechanical properties in regions as small as a few nanometers.

Typical applications of AFM include 3D topography of integrated circuits (IC), roughness measurements for various materials, analysis of microscopic phase distribution in polymers, mechanical and physical property measurements for thin films, imaging magnetic domains on digital storage media, imaging of submicron phases in metals, defect imaging in IC failure analysis, microscopic imaging of fragile biological samples, metrology for compact disk stampers. No sample preparation is typically required. Samples can be imaged in air or liquid. Sample height is limited to about 4 cm. Areas up to 20 cm in diameter can be fully traversed without repositioning. Larger samples can be fixed for imaging within a limited area. Total surface roughness in the image area should not exceed 1 μm.

Scanning electron microscopy (SEM) is a method for high-resolution imaging of surfaces [1,2]. It uses electrons for imaging and its advantages over LM include much higher magnification (>100,000x) and greater depth of field up to 100 times that of light microscopy. Qualitative and quantitative chemical analysis information is also obtained using an energy dispersive x-ray spectrometer (EDS) with the SEM (see more details about EDS later in this chapter). The SEM generates a beam of incident electrons in an electron column above the sample chamber. The electrons are produced by a thermal emission source, such as a heated tungsten filament, or by a field emission cathode. The energy of the incident electrons can be as low as 100 eV or as high as 30 keV depending on the evaluation objectives. The electrons are focused into a small beam by a series of electromagnetic lenses in the SEM column. Scanning coils near the end of the column direct and position the focused beam onto the sample surface. The electron beam is scanned in a raster pattern over the surface for imaging. The beam can also be focused at a single point or scanned along a line for x-ray analysis. The beam can be focused to a final probe diameter as small as about 10 Å. The incident electrons cause electrons to be emitted from the sample due to elastic and inelastic scattering events within the sample's surface and near-surface material. High-energy electrons that are ejected by an elastic collision of an incident electron, typically with a sample atom's nucleus, are referred to as backscattered electrons. The energy of backscattered electrons will be comparable to that of the incident electrons. Emitted lower-energy electrons resulting from inelastic scattering are called secondary electrons. Secondary electrons can be formed by collisions with the nucleus where substantial energy loss occurs or by the ejection of loosely bound electrons from the sample atoms. The energy of secondary electrons is typically 50 eV or less. To create an SEM image, the incident electron beam is scanned in a raster pattern across the sample's surface. The emitted electrons are detected for each position in the scanned area by an electron detector. The intensity of the emitted electron signal is displayed as brightness on a cathode ray tube. By synchronizing the cathode ray tube scan to that of the scan of the incident electron beam, the cathode ray tube display represents the morphology of the sample surface area scanned by the beam.

The SEM column and sample chamber are at a moderate vacuum to allow the electrons to travel freely from the electron beam source to the sample and then to the detectors. High-resolution imaging is done with the chamber at higher vacuum, typically from 10^{-5} to 10^{-7} Torr. Imaging of nonconductive, volatile, and vacuum-sensitive samples can be performed at higher pressures.

Secondary electron imaging mode provides high-resolution imaging of fine surface morphology with SEM. Inelastic electron scattering caused by the interaction between the sample's electrons and the incident electrons results in the emission of low-energy electrons from near the sample's surface. The topography of surface features influences the number of electrons that reach the secondary electron detector from any point on the scanned surface. This local variation in electron intensity creates the image contrast that reveals the surface morphology. The secondary electron image resolution for an ideal sample is about 3.5 nm for a tungsten-filament electron source SEM or 1.5 nm for field emission SEM.

Backscatter electron imaging mode provides image contrast as a function of elemental composition, as well as, surface topography. Backscattered electrons are produced by the elastic interactions between the sample and the incident electron beam. These high-energy electrons can escape from much deeper than secondary electrons, so surface topography is not as accurately resolved as for secondary electron imaging. The production efficiency for backscattered electrons is proportional to the sample material's mean atomic number, which results in image contrast as a function of composition, i.e., higher atomic number material appears brighter than low atomic number material in a backscattered electron image. The optimum resolution for backscattered electron imaging is about 5.5 nm.

Variable pressure SEM mode is used when the sample is not electrically-conductive to allow incident electrons to be conducted away from the sample surface to the ground. If electrons accumulate on a nonconductive surface, the charge buildup causes a divergence of the electron beam and degrades the SEM image. In variable-pressure SEM, some air is allowed into the sample chamber, and the interaction between the electron beam and the air molecules create a cloud of positive ions around the electron beam. These ions will neutralize the negative charge from electrons collecting on the surface of a nonconductive material. SEM imaging can be performed on a nonconductive sample when the chamber pressure is maintained at a level where most of the electrons reach the sample surface, but there are enough gas molecules to ionize and neutralize the charge. Variable pressure SEM is also valuable for examination of biological samples that are not compatible with high vacuum.

Quantification is possible with SEM because the image magnification is calibrated against a reference standard. Lateral feature dimensions can be readily quantified to an accuracy of less than 0.1 μm. Computer analysis of images can quantify area or volume fractions and particle shapes and sizes.

SEM finds applications in microscopic feature measurement, fracture characterization, microstructure studies, thin coating evaluations, surface contamination examination, IC failure analysis, imaging of biological samples, etc. In a large-chamber SEM, samples up to 200 mm in diameter can be readily accommodated. Larger samples, up to 300 mm can be loaded with limited stage movement. Sample height is typically limited to ~ 50 mm. Backscattered electron imaging can be performed on conductive or nonconductive samples. For secondary electron imaging, samples must be electrically conductive. Nonconductive materials can be evaporative coated with a thin film of carbon, gold or other conductive material to obtain conductivity without significantly affecting observed surface morphology but it affects the chemistry of the studied surface. Samples must be compatible with at least a moderate vacuum. For high-resolution secondary electron imaging, the sample environment is at a pressure of 1 x 10-5 Torr or less. The pressure can be adjusted up to about 2 Torr for vacuum sensitive samples.

Transmission electron microscopy (TEM) uses a high energy beam of electrons that is shone through a very thin sample and the interactions between the electrons, and the atoms can be used to observe features such as the crystal structure and features in the structure like dislocations and grain boundaries [3]. The TEM microscope operates on the same basic principles as the LM but uses electrons instead of light. Because the wavelength of electrons is much smaller than that of light, the optimal resolution attainable for TEM images is many orders of magnitude better than that from a LM. Thus, TEM can reveal the finest details of the internal structure - in some cases as small as individual atoms.

Imaging mode: the beam of electrons from the electron gun is focused into a small, thin, coherent beam by the use of a condenser lens. This beam is restricted by the condenser aperture, which excludes high angle electrons. The beam then strikes the specimen and parts of it are transmitted depending upon the thickness and electron transparency of the specimen. This transmitted portion is focused by the objective lens into an image on a phosphor screen or charge coupled device (CCD) camera. Optional objective apertures can be used to enhance the contrast by blocking out high-angle diffracted electrons. The image then passed down the column through the intermediate and projector lenses where it is enlarged. The image strikes the phosphor screen and light is generated, allowing the user to see the image. The darker areas of the image represent those areas of the sample where fewer electrons are transmitted through while the lighter areas of the image represent those areas of the sample where more electrons were transmitted through.

Diffraction mode: as the electrons pass through the sample, they are scattered by the electrostatic potential set up by the constituent elements in the specimen. After passing through the specimen they pass through the electromagnetic objective lens which focuses all the electrons scattered from one point of the specimen into one point in the image plane. This is the back focal plane of the objective lens and is where the diffraction pattern is formed.

An advantage of the TEM technique is the high resolution one can obtain for sample imaging but a disadvantage is the need for very special specimen preparation. A TEM specimen must be thin enough to transmit sufficient electrons to form an image with minimum energy loss. Therefore specimen preparation is an important aspect of the TEM analysis. For most electronic materials, a common sequence of preparation techniques is ultrasonic disk cutting, dimpling, and ion-milling. Dimpling is a preparation technique that produces a specimen with a thinned central area and an outer rim of sufficient thickness to permit ease of handling. Ion milling is traditionally the final form of specimen preparation. In this process, charged argon ions are accelerated to the specimen surface by the application of high voltage. The ion impingement upon the specimen surface removes material as a result of momentum transfer.

TEM is a very powerful tool for material science. TEM can be used to study the growth of layers, their composition and defects in semiconductors. High resolution can be used to analyze the quality, shape, size and density of quantum wells, wires and dots. A chemical analysis can also be performed.

Energy dispersive X-ray spectroscopy (EDS or EDX) is a chemical microanalysis technique used in conjunction with SEM [1,2]. The EDS technique detects x-rays emitted from the sample during bombardment by an electron beam to characterize the elemental composition of the analyzed volume. Features or phases as small as 1 μm or less can be analyzed. When the sample is bombarded by the SEM's electron beam, electrons are ejected from the atoms comprising the sample's surface. The resulting electron vacancies are filled by electrons from a higher state, and an x-ray is emitted to balance the energy difference between the two electrons' states. The x-ray energy is characteristic of the element from which it was emitted. The EDS x-ray detector measures the relative abundance of emitted x-rays versus their energy. When an incident x-ray strikes the detector, it creates a charge pulse that is proportional to the energy of the x-ray. The charge pulse is converted to a voltage pulse (which remains proportional to the x-ray energy) by a charge-sensitive preamplifier. The signal is then sent to a multichannel analyzer where the pulses are sorted by voltage. The energy, as determined from the voltage measurement, for each incident x-ray is sent to a computer for display and further data evaluation. The spectrum of x-ray energy versus counts is evaluated to determine the elemental composition of the sampled volume.

For qualitative analysis the sample x-ray energy values from the EDS spectrum are compared with known characteristic x-ray energy values to determine the presence of an element in the sample. Elements with atomic numbers ranging from that of beryllium to uranium can be detected. The minimum detection limits vary from approximately 0.1 to a few atom percent, depending on the element and the sample matrix. Quantitative results can be obtained from the relative x-ray counts at the characteristic energy levels for the sample constituents. Semi-quantitative results are readily available without standards by using mathematical corrections based on analysis parameters and the sample composition. The accuracy of standardless analysis depends on the sample composition. Greater accuracy is obtained using known standards with similar structure and composition to that of the unknown sample.

Elemental mapping is performed as the characteristic x-ray intensity is measured relative to lateral position on the sample. Variations in x-ray intensity at any characteristic energy value indicate the relative concentration for the applicable element across the surface. One or more maps are recorded simultaneously using image brightness intensity as a function of the local relative concentration of the element(s) present. Lateral resolution of about 1 μm is possible.

For line profile analysis with the EDS the electron beam is scanned along a preselected line across the sample while x-rays are detected for discrete positions along the line. Analysis of the x-ray energy spectrum at each position provides plots of the relative elemental concentration for each element versus position along the line.

Typical applications of the EDS technique include foreign material analysis, corrosion evaluation, coating composition analysis, rapid material alloy identification, small component material analysis, phase identification and distribution. Maximum sample sizes are as described for the SEM analysis and they must also be compatible with a moderate vacuum atmosphere (pressures of 2 Torr or less).

X-ray photoelectron spectroscopy (XPS), also known as electron spectroscopy for chemical analysis, is an analysis technique used to obtain chemical information about the surfaces of solid materials [1,2]. Both composition and the chemical state of surface constituents can be determined. Insulators and conductors can easily be analyzed in surface areas from a few microns to a few millimeters across. The sample is placed in an ultrahigh vacuum environment and exposed to a low-energy, monochromatic x-ray source. The incident x-rays cause the ejection of core-level electrons from sample atoms. The energy of a photoemitted core electron is a function of its binding energy and is characteristic of the element from which it was emitted. Energy analysis of the emitted photoelectrons is the primary data used for XPS. When the core electron is ejected by the incident x-ray, an outer electron fills the core hole. The energy of this transition is balanced by the emission of an Auger electron or a characteristic x-ray. Analysis of Auger electrons can be used in XPS, in addition to emitted photoelectrons. The photoelectrons and Auger electrons emitted from the sample are detected by an electron energy analyzer, and their energy is determined as a function of their velocity entering the detector. By counting the number of photoelectrons and Auger electrons as a function of their energy, a spectrum representing the surface composition is obtained. The energy corresponding to each peak is characteristic of an element present in the sampled volume. The area under a peak in the spectrum is a measure of the relative amount of the element represented by that peak. The peak shape and precise position indicates the chemical state for the element. XPS is a surface sensitive technique because only those electrons generated near the surface escape and are detected. The photoelectrons of interest have relatively low kinetic energy. Due to inelastic collisions within the sample's atomic structure, photoelectrons originating more than 20 to 50 Å below the surface cannot escape with sufficient energy to be detected.

Energy peaks in a survey scan mode identify the elemental composition of the uppermost 20 to 50 Å of the analyzed surface. All elements, except hydrogen and helium, are detected. Detection limits are approximately 0.1 atom percent for most elements.

High resolution multiplex scan mode evaluates the chemical state(s) of each element through its core electron binding energies. Precise determination of binding energies is made through the use of curve fitting routines applied to the peaks in the multiplex scan. Shifts in the binding energy can result from the atom's oxidation state, chemical bonds, or crystal structure. A NIST database is available to identify binding energies with specific compounds.

Quantification is obtained based on the determination of the concentrations of the elements identified in the survey scan by integrating the area under a characteristic peak for each element. Sensitivity factors are applied to the peak area values to determine the elemental concentration. The elemental composition can be measured as a function of depth into the sample by alternating Auger electron spectroscopy (AES) analysis with ion sputtering to remove material from the sample surface (for AES information see further in the section). Depth resolution of <100 Å is possible. The relative concentration of one or more elements is determined as a function of lateral position on the sample surface in the mapping mode. An image is obtained where brightness indicates the element concentration.

Typical applications include analysis of thin film contamination, evaluation of adhesion failures, measurement of elemental composition of insulating materials (e.g., polymers, glasses), identification of the chemical state of surface films (e.g. metal or oxide), quantitative elemental depth profiling of insulators. Sample size cannot exceed 25 mm in any lateral direction and height should not exceed 12 mm. Sample must be compatible with an ultrahigh vacuum environment ($>10^{-9}$ Torr).

Auger electron spectroscopy (AES) provides information about the chemical composition of the outermost material comprising a solid surface or interface [1,2]. The principal advantages of AES over other surface analysis methods are excellent spatial resolution (< 1 μm), surface sensitivity (~20 Å), and detection of light elements. Detection limits for most elements range from about 0.01 to 0.1 at%. AES uses a primary electron beam to excite the sample surface. When an inner-shell electron is ejected from a sample atom by the interaction with a primary electron, an electron from an outer shell fills the vacancy. To compensate for the energy change from this transition, an Auger electron or an x-ray is emitted. For light elements, the probability is greatest for the emission of an Auger electron, which accounts for the light element sensitivity for this technique. The energy of the emitted Auger electron is characteristic of the element from which it was emitted. Detection and energy analysis of the emitted Auger electrons produces a spectrum of Auger electron energy versus the relative abundance of electrons. Peaks in the spectrum identify the elemental composition of the sample surface. In some cases, the chemical state of the surface atoms can also be determined from energy shifts and peak shapes. Auger electrons have relatively low kinetic energy, which limits their escape depth. Any Auger electrons emitted from an interaction below the surface will lose energy through additional scattering reactions along its path to the surface. Auger electrons emitted at a depth greater than about 2-3 nm will not have sufficient energy to escape the surface and reach the detector. Thus, the analysis volume for AES extends only to a depth of about 2 nm. Analysis depth is not affected by the energy of the primary electron energy. The AES instruments are equipped for SEM to facilitate location of selected analysis areas, and micrographs of the sample surface can be obtained. The sample chamber is maintained at ultrahigh vacuum to minimize interception of the Auger electrons by gas molecules between the sample and the detector. Some instruments include special stages for fracturing samples to examine interfaces that have been freshly exposed within the vacuum chamber. A computer is used for acquisition, analysis, and display of the AES data.

The position of the peaks in the AES spectrum obtained in a survey scan mode identifies the elemental composition of the uppermost 20 Å of the analyzed surface.

Multiplex Scan: a higher resolution analysis of the Auger spectrum in the region around a characteristic peak is used for determination of the atomic concentration of the elements identified in the survey scans and in some cases chemical state information. The AES analysis results can be quantified without standards by using the area under the peaks in the AES spectrum and corrections based on elemental sensitivity factors. Mapping and Line scans are imaging techniques that measure the lateral distribution of elements on the surface. The electron beam is scanned across the sample surface, either along a fixed line (line scan) or across a given area (mapping) while the AES signal is analyzed for specific energy channels. The AES signal intensity is a function of the relative concentration of the element(s) corresponding to the selected energy channel(s). Spatial resolution is approximately 0.3 μm. In the depth profile mode material is removed from the surface by sputtering with an energetic ion beam concurrent with successive AES analyses. This process measures the elemental distribution as a function of depth into the sample and depth resolution of < 100 Å is possible.

Typical applications of AES include microscopic particle identification, passive oxide film thickness, contamination on integrated circuits, quantification of light element surface films, mapping spatial distribution of surface constituents, etc. Sample requirements are the same as for the XPS analysis. In addition, they must be conductive or area of interest must be properly grounded. Insulating samples, including thick insulating films (>300 Å), cannot be analyzed.

Fourier transform infrared spectroscopy (FTIR) is an analytical technique used to identify organic (and in some cases inorganic) materials [1,2]. This technique measures the absorption of infrared radiation by the sample material versus wavelength. The infrared absorption bands identify molecular components and structures. When a material is irradiated with infrared radiation, absorbed IR radiation usually excites molecules into a higher vibrational state. The wavelength of light absorbed by a particular molecule is a function of the energy difference between the at-rest and excited vibrational states. The wavelengths that are absorbed by the sample are characteristic of its molecular structure. The FTIR spectrometer uses an interferometer to modulate the wavelength from a broadband infrared source. A detector measures the intensity of transmitted or reflected light as a function of its wavelength. The signal obtained from the detector is an interferogram, which must be analyzed with a computer using Fourier transforms to obtain a single-beam infrared spectrum. The FTIR spectra are usually presented as plots of intensity versus wavenumber (in cm^{-1}). Wavenumber is the reciprocal of the wavelength. The intensity can be plotted as the percentage of light transmittance or absorbance at each wavenumber.

Qualitative material identification mode: to identify the material being analyzed, the unknown IR absorption spectrum is compared with standard spectra in computer databases or with a spectrum obtained from a known material. Spectrum matches identify the polymer or other constituent(s) in the sample. Absorption bands in the range of 4000 - 1500 wavenumbers are typically due to functional groups (e.g., -OH, C=O, N-H, CH_3, etc.). The region from 1500-400 wavenumbers is referred to as the fingerprint region. Absorption bands in this region are generally due to intramolecular phenomena and are highly specific to each material. The specificity of these bands allows computerized data searches within reference libraries to identify a material.

Quantitative concentration of a compound can be determined from the area under the curve in characteristic regions of the IR spectrum. Concentration calibration is obtained by establishing a standard curve from spectra for known concentrations.

Typical applications of FTIR include identification of foreign materials, particulates, fibers, residues, identification of bulk material compounds, identification of constituents in multilayered materials, quantification of silicone, esters, etc., as contamination on various materials, etc. Sample requirements vary depending on the sample form and instrument. Samples may be in liquid, solid or gaseous form. When using a microscope attachment on the spectrometer, the analysis area can be as small as 10 μm. Thin organic films on a reflective surface (e.g., gold) can be analyzed in situ using the microscope's reflectance mode. The outer 1-10 μm of a material can be analyzed using attenuated total reflectance attachment.

Raman spectroscopy is a form of molecular spectroscopy that involves the scattering of electromagnetic radiation by atoms or molecules [4,5]. It probes the vibrational, rotational, and other low-frequency modes of molecules; the Raman signal is observed as inelastic scattered light. When electromagnetic radiation is scattered by a molecule or by a crystal, one photon of the incident radiation is annihilated and at the same time one photon of the scattered radiation is created. The scattering mechanisms can be classified on the basis of the difference between the energies of the incident and scattered photons. If the energy of the incident photon is equal to that of the scattered one, the process is called Rayleigh scattering. If the energy of the incident photon is different to that of the scattered one, the process is called Raman scattering. The Raman scattering (or Raman effect) was discovered in 1928 by V.C. Raman. Thanks to the development of the laser light, it is now a very important technique in the study of matter structure. "If the substance being studied is illuminated by monocromatic light of frequency f_0, the spectra of the scattered light consists of a strong line (the exciting line) of the same frequency as the incident illumination together with weaker lines on either side shifted from the strong line by frequencies ranging from a few to about 3500 cm^{-1}. The lines of frequency less than the exciting lines are called Stokes lines, the others anti-Stokes lines. The Stokes lines are more intense than the anti-Stokes lines.

In crystalline solids, the Raman effect deals with phonons, instead of molecular vibration. A phonon is Raman-active only if the first derivative of the polarizability with respect the vibrational normal coordinate has a non-zero value, and this in turns depends on the crystal symmetry. A phonon can be either IR or Raman active only in crystals without a center of inversion. Performing measurements in controlled polarization configurations makes it possible to obtain information about the symmetry of the crystalline lattice. The Raman signal is very weak: only one photon in 10^7 gives rise to the Raman effect. The Raman spectra are usually plotted in intensity vs. the difference in wavenumber between the incident beam and the scattered light, and so the peaks are in correspondence to the phonon frequency. Due to the small wave-vector of the optical photons, the phonons involved in the Raman scattering of crystalline solids have (for the wave-vector conservation law) a very small momentum compared with the Brillouin zone, so only the zone-center phonons participate to the Raman scattering. In disordered solids this rule is no longer valid, and all the phonons contribute to the Raman spectrum, leading to large bands that reflect the vibrational density of states. In nano-crystalline materials, the situation is intermediate between ideal crystals and amorphous, and the Raman spectrum displays the crystalline Raman features broadened and shifted by the phonon confinement. By using an adequate model, one can estimate the size of the nanocrystals. From the band-shifts and the presence of "forbidden" peaks, using the Raman spectroscopy also make it possible to obtain information on the disorder and strains present in the crystalline lattice. Infrared spectroscopy and Raman spectroscopy are complementary techniques. For highly symmetric polyatomic molecules possessing a center of inversion (such as benzene) it is observed that bands that are active in the IR spectrum are not active in the Raman spectrum (and vice-versa). In molecules with little or no symmetry, modes are likely to be active in both infrared and Raman spectroscopy. Thus the two techniques used together are very useful for sample structure identification.

There are number of advantages of the Raman spectroscopy, for example it is a nondestructive technique and typically requires little to no sample preparation. The Raman analysis also can be performed directly through transparent containers, including plastic bags, glasses, jars, cuvettes, and so on. Furthermore, the Raman technique can be used for both qualitative and quantitative analysis and is highly selective, meaning that it is able to differentiate molecules in chemical species that are very similar. Raman also has the advantage of quick analysis times. A typical analysis can take just a few seconds, and unlike FTIR spectroscopy Raman is insensitive to aqueous absorption bands.

Raman spectroscopy finds applications in pharmaceutical, food and agriculture industries, forensic and carbon analysis, mineralogy, art, biomedical area, etc. An advantage of Raman spectroscopy is that it is a nondestructive test method requiring little to no sample preparation. Other advantages of Raman spectroscopy include selectivity and flexibility for both qualitative and quantitative analysis. It is possible to lower the laser power to very low levels to ensure that samples are not burnt or damaged in any way and in order to see sufficient Raman scattering coming back to the detector, sensitive CCDs are used.

X-ray diffraction (XRD) is a non-destructive and rapid analytical technique primarily used for phase identification of a crystalline material and can provide information on unit cell dimensions [6]. In 1912, Max von Laue, a German physicist, discovered that x-rays could be diffracted, or scattered, in an orderly way by the orderly array of atoms in a crystal. That is, crystals can be used as three-dimensional 'diffraction gratings' for x-rays. The phenomenon of x-ray diffraction from crystals is used both to analyze x-rays of unknown wavelength using a crystal whose atomic structure is known, and to determine, using x-rays of known wavelength, the atomic structure of crystals.

XRD analysis is based on constructive interference of monochromatic X-rays and a crystalline sample: The X-rays are generated by a cathode ray tube, filtered to produce monochromatic radiation, collimated to concentrate, and directed toward the sample. The interaction of the incident rays with the sample produces constructive interference (and a diffracted ray) when conditions satisfy Bragg's Law ($n\lambda = 2d.\sin\theta$). This law relates the wavelength of electromagnetic radiation λ to the diffraction angle θ and the lattice spacing in a crystalline sample d.

The characteristic x-ray diffraction pattern generated in a typical XRD analysis provides a unique "fingerprint" of the crystals present in the sample. When properly interpreted, by comparison with standard reference patterns and measurements, this fingerprint allows identification of the crystalline form.

Knowledge about crystallinity is highly relevant, as a crystalline form is usually preferred in development. In contrast to amorphous material, a crystal has well-defined properties. The result from an XRD analysis is a diffractogram showing the intensity as a function of the diffraction angles. Positive ID of a material using XRD analysis is based on accordance between the diffraction angles of a reference material and the sample in question.

XRD analysis is a unique method in determination of crystallinity of a compound. The method has been traditionally used for identification of crystalline materials, phase identification, quantitative analysis and the determination of structure imperfections. In recent years, applications have been extended to new areas, such as the determination of crystal structures and the extraction of three-dimensional microstructural properties. The method is applied to data collected under ambient conditions, but in situ diffraction as a function of external constraints (temperature, pressure, stress, electric field, atmosphere, etc.) is important for the interpretation of solid state transformations and materials behavior. Various kinds of micro- and nanocrystalline materials can be characterized from X-ray powder diffraction, including inorganics, organics, drugs, minerals, zeolites, catalysts, metals and ceramics. The physical states of the materials can be loose powders, thin films, polycrystalline and bulk materials. For most applications, the amount of information which is possible to extract depends on the nature of the sample microstructure (crystallinity, structure imperfections, crystallite size, and texture), the complexity of the crystal structure (number of atoms in the asymmetric unit cell, unit cell volume) and the quality of the experimental data (instrument performances, counting statistics). Standard reference materials are available for instrument calibration, evaluation of instrument characteristics, quantitative analysis, intensity sensitivity, etc.

Phase identification is traditionally based on a comparison of observed data with interplanar spacing d and relative intensities I compiled for crystalline materials. The Powder Diffraction File, edited by the International Centre for Diffraction Data (USA), contains powder data for more than 130 000 substances.

Quantitative phase analysis involves the determination of the amounts of different phases present in a multi-component mixture. The powder method is widely used to determine the abundance of distinct crystalline phases. Errors due to residual preferred orientation effects are reduced by considering the whole pattern.

Microstructural imperfections, such as lattice distortions, stacking and twin faults, dislocations, the small size of crystallites (i.e. nanoscale domains over which diffraction is coherent) and crystallite size distributions, are usually extracted from the integral breadth or a Fourier analysis of individual diffraction lines. Crystallite sizes well determinable by line broadening analysis are in the range of 20-1500 Å. Crystallite shape anisotropy has been determined in strain-free nanocrystalline materials. Lattice distortion (microstrain/stress) represents variable displacements of atoms from their sites in the idealized crystal structure. Anisotropic microstrains have been successfully interpreted by the presence of (specific) dislocation distributions in various materials. Materials behavior is largely determined by the presence of macro- (engineering stresses) and microstresses (e.g. thin films). Powder diffraction is the only experimental method available for description of both micro- and macro-stresses.

Time- and temperature-dependent XRD includes the measurement of series of diffraction patterns as a function of time and temperature. The time required for collecting data decreases considerably with the availability of fast detectors, such as position sensitive detectors, and high brightness of the X-ray source. Limitations with conventional X-rays can be overcome with synchrotron and neutron diffraction facilities. Equipment has been developed for in situ application of heating, pressure and tensile testing. In principle, line profile parameters can be extracted from each pattern (peak position and integrated intensity, breadth and line shape) and interpreted in structural and microstructural terms.

2. Coherence Probe Microscopy Development for Imaging Thick and Complex Biomaterial Layers

The process by which organisms in nature make minerals is known as biomineralization. In recent decades it has been intensively studied due to its importance in a wide range of biological events, starting from the formation of bones, teeth, cartilage, shells, corals, and extending to pathological biomineralization, i.e. the formation of kidney stones, dental calculi, arteriosclerosis, osteogenesis, etc. HA ($Ca_{10}(PO_4)_6(OH)_2$) is an inorganic material that is chemically similar to the mineral component of bones, teeth and hard tissues in mammals. It is one of the few materials that are classed as bioactive, meaning that it supports bone ingrowth and osseointegration when implanted into a living body. Due to its excellent bioactivity and biocompatibility properties, HA is used *in-vivo* as a coating for hard tissue implants in the human body (orthopaedic, dental), bone fillers, in bone repair and bone augmentation, for drug delivery, etc.

In the work presented here, HA layers were prepared by using a biomimetic approach that mimics the natural process of mineral formation [7-9]. It involves the formation of inorganic HA from a supersaturated SBF, an aqueous solution that resembles the inorganic composition and concentration of human blood plasma (Table 2). Stainless steel (metal widely used in medicine and dentistry for hard tissue implants), silicon (Si: well studied and widely used semiconductor material in microelectronics, now used for the fabrication of biosensors) and silica glass (insulating material, the glassy form of crystalline quartz, investigated for its application in biosensors) are used as substrates. Different materials are used to cover a wide range of surfaces for studying the process of biomineralization. Another goal is to connect micro- and optoelectronics with living tissue for the creation of biosensor devices.

Table 2. Ion concentrations (mM) of human blood plasma, Kokubo's SBF [7] and the solution used in the experiments described in this book (m-SBF)

	Ion concentrations (mM)						
	Na^+	K^+	Mg^{2+}	Ca^{2+}	Cl^-	HCO_3^-	HPO_4^{2-}
Human plasma	142.0	5.0	1.5	2.5	103.0	27.0	1.0
SBF	142.0	5.0	1.5	2.5	148.8	4.2	1.0
m-SBF	142.0	5.0	1.5	3.75	148.8	4.2	1.5

For the first time this solution has been studied by Kokubo and colleagues [7] who by varying the ion composition, concentrations and pH reached the conclusion that SBF with specified parameters resembling the human blood plasma is most appropriate for obtaining biologically active apatite mimicking the natural apatite. Advantage of this method is not only its resemblance to the process of mineral formation in nature but also the possibility to add biologically active molecules in the solution, polymeric substances, therapeutic substances, and substances that promote cellular growth.

Our experiments are based on the well-established Kokubo's SBF [7] for which it is known that prolonged immersion of the samples is necessary for obtaining precipitation. The original composition of Kokubo was prepared and modified with 1.5 times higher Ca and PO_4 concentrations (solution named m-SBF in Table 2), thus obtaining faster visible precipitation on the samples surface [8,9]. The preparation of the m-SBF involves preparation of two base solutions of reagent grade chemicals. The first contains NaCl (136.8 mM), KCl (3 mM), $MgCl_2 \cdot 6H_2O$ (1.5 mM), $CaCl_2 \cdot 6H_2O$ (3.75 mM), dissolved in 1000 ml doubly distilled water. The second contains $Na_2SO_4 \cdot 10H_2O$ (0.5 mM), $NaHCO_3$ (4.2 mM), $K_2HPO_4 \cdot 3H_2O$ (1.5 mM), also dissolved in 1000 ml doubly distilled water. Both base solutions do not yield precipitation when stored separately. Buffering of each of them at 7.4 with TRIS buffer (tris-hydroxymethyl-aminomethane, $(CH_2OH)_3CNH_2$, when pH < 7.4) or hydrochloric acid (HCl, when pH > 7.4) was applied. Before the experiments, equal parts of the two solutions were mixed to form the final m-SBF.

The experimental set up developed for the growth of HA layers from the prepared m-SBF on different solid surfaces is shown in Fig. 1. It presents an open deposition system and include container with the m-SBF, whose temperature during the experiment is maintained at 37 °C. The substrates are fixed in the solution on a holder in horizontal or vertical position. Layer growth is carried out by a prolonged soaking of the samples in 400 ml total volume of the supersaturated solution, equivalent to 40 mm aqueous layer over the substrate surfaces (20-40 ml/sample) with different soaking duration depending on the experiment: from 2 to 108 hours. When the samples are kept in the solution for more than 24 hours, refreshment of the solution is applied which means that 100 ml used solution is taken out and replaced by 100 ml fresh m-SBF for maintaining the ion concentrations and supersaturation until the end of the experiment. Concentrations of Ca and P ions are measured every day by inductively coupled plasma mass spectrometry and ion chromatography.

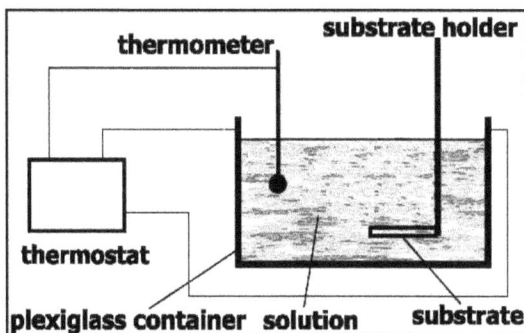

Figure 1. Experimental set up for growth of HA from m-SBF.

The technique used for imaging of the obtained thick and complex HA layers is based on *white light scanning interferometry* (WLSI), which is also known as *coherence probe microscopy* (CPM). In this chapter, the classical use of CPM in surface roughness measurement and different ideas for new measurement modes possible with this technique are described, modes that are particularly useful in thick layer characterization.

WLSI has been widely used in microelectronics, optics and medicine since the end of the 1980's for the characterization of integrated circuits and semiconductor materials [10-15], microsystems, step heights and discontinuities [16,17], biological tissues (optical coherence tomography imaging) [18,19], optical components [20] and rough surfaces [21,22]. However, to the best of our knowledge, it has not been applied to the characterization of biomimetic layers such as HA. CPM is a good solution for observing and measuring the topography of large areas and deep roughness, and for profiling difficult surfaces such as those that are fragile or contain buried interfaces. CPM is an extremely powerful tool for carrying out optical measurements rapidly, without contact, and with high precision. Depending on the sample depth, the topography and roughness, a surface can be measured in a matter of seconds or minutes. Because of the nature of the optical probe used in CPM, the technique is referred to as low coherence optical probe microscopy. The CPM technique is complementary to other surface measurement techniques such as stylus, confocal microscopy, AFM, SEM and TEM which when used together are very useful because they can give an overall idea for the complex HA layers. Classical stylus profilometry is usually used on hard materials (metals, glasses, Si, etc.) because it requires physical contact with the surface; the tip force is too high to work on the HA layer without digging into it, and the technique has limited lateral resolution. AFM is very useful for probing nanometer features but it is difficult to measure precisely the topography and profile of thick and rough layers in the micrometer range such as those used in this work, without causing damage either to the tip or to the surface layer due to its limited vertical range. Additionally, AFM can only be performed on small areas. Qualitative information could be obtained by SEM but at high cost, special sample preparation (coating with carbon, cross-sectioning) and high time-consumption. SEM cannot easily be used to give quantitative topographic information detailed enough for these sorts of layers. Since topographical data of biological relevance is required over a large range from nanometers to micrometers, it is therefore important to combine the results from AFM and SEM with those from other techniques such as CPM.

The CPM technique is based on the principle of white light interferometry used in a reflection microscope in which the wavefront reflected from a flat reference mirror is compared with that from the surface of the sample measured. An interference objective (Mirau or Linnik type, Fig. 2) is used to give a series of interference fringes superimposed on an image of the surface of the sample acquired by a CCD camera. The intensity profile of the white light fringes is used as a virtual probe plane that is scanned along the depth of the sample (i.e. the optical axis of the microscope). The peak of the envelope of the interference fringes corresponds to the position of the surface at that point. By using specialized algorithms for detecting the peak of the envelope at each pixel in the image, the complete topography of the layer can be measured [10-13]. The technique has nanometric axial resolution (1 to 80 nm depending on the conditions and algorithms used), submicron lateral resolution, a large depth of field and a high axial dynamic range (up to 100 μm with our system) [14,15,21,22]. The uncertainty in the measurement of surface height can be as low as a few nm on simple flat structures, but this value can increase in certain conditions by as much as $\lambda/2$ on more complex structures.

The results are displayed in a grayscale level image in which the grayscale corresponds to the height of the layer. Three types of images are obtained to analyze the surfaces under study: grayscale (2D image), reflection and 3D. The grayscale image (height view) provides quantitative data that can be analyzed to give statistical measurements of line profiles of the surface and roughness values and a 3D view to give a qualitative impression of the surface topography. The reflection image also provides a qualitative view of the surface; its extended depth of field covering the whole depth of the sample roughness provides supplementary information concerning the homogeneity of the surface, indicating variations in reflectivity that can be due to the presence of different materials or changes in the surface angle. An additional mode of measurement that we have developed is single point Z scanning in order to analyze the fringe signal at a given point in the image, leading to useful information concerning the presence of transparent layers and buried interfaces.

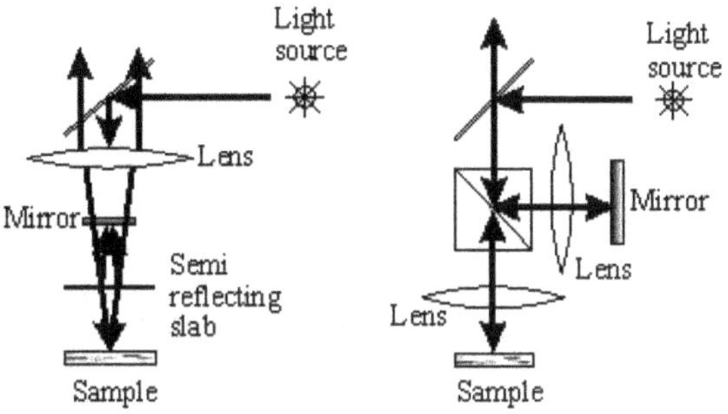

Mirau: x10, x20, x40   Linnik: x50, x100

Figure 2. Details of the two types of interference objectives used in this work and the typical magnifications associated with each.

Substrates (8 x 8 mm in size) were prepared from AISI 316 stainless steel, n-type Si with (100) orientation and silica glass Herasil. Subsequent ion implantation of Ca and P with parameters described in [8] was carried out for selective surface modification of the materials and to induce nucleation points on the surface for facilitating the HA growth. A group of implanted samples was subsequently subjected to thermal treatment in air (600°C, 1 hour) in order to yield the formation of Ca and P-based oxides [8]. For the biomimetic growth of the HA layers, supersaturated SBF was prepared by dissolving reagent-grade chemicals in doubly distilled water, and the modified and control (polished) samples were immersed for 6 days under natural conditions (37 °C, pH 7.4). As a result, CO_3-containing HA layers were grown [8,9,22].

Surface topography was observed with an optical reflection microscope (Axioskop, Zeiss Jena, magnification of 20x), AFM scanning probe and optical microscope (DualScope C-21, DME, tapping mode and Park XE 70, non-contact mode), Zygo NewView 7200 interference microscope (x50 Mirau objective, digital b&w camera with 640x480 pixels, Zygo proprietary image analysis software) and CPM (Leitz Linnik microscope based system, magnification of 50x). The layer morphology was observed by SEM (DSM 962, Carl Zeiss, Oberkochen, magnifications of 500x and 5000x). Rutherford backscattering (RBS, 4 MV Van de Graaff accelerator, InESS, ^{4}He incident beam, 2 MeV) was employed to identify the transparent layer observed with the Leitz microscope on the silicon samples underneath the HA layers. The SAM program (Simulation for Analysis of Materials developed at the ICube, University of Strasbourg) was used for simulations of RBS data by entering the experimental conditions and the sample composition.

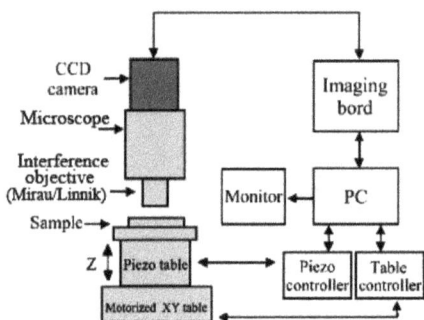

Figure 3. Layout of the CPM measurement system.

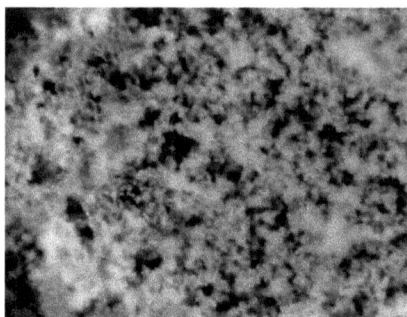

Figure 4. LM image of thick heterogeneous HA layer consisting of white crystals (image size 320 x 240 μm).

A layout of the CPM measurement system is shown in Fig. 3. It is based on a Leitz microscope equipped with a Linnik interferometer with a 50x objective (NA 0.85) for producing white light interference fringes superimposed on an image of the surface. An incandescent lamp (λ_{centre} = 610 nm) was used for the illumination with white light having a spectral range of 350-1100 nm and resulting in a theoretical lateral resolution of 0.45 μm. The sample was mounted on a piezo-controlled table having a linear variable differential transformer (LVDT) position control with a vertical resolution of 10 nm over a dynamic range of 100 μm. The piezo-table was mounted on precision stainless steel manual XY tables. A CCD camera (Sony XC-75 CE) and a standard 8-bit depth imaging board were used to produce a series of digitized images, which were stored and processed on a PC (Pentium III, 1 GHz). Graphical programming was used to write the software for controlling the system and for carrying out image processing and analysis of the results. Various algorithms have been developed to process the recorded images [10-13]. CPM measurements were performed on each of the samples over an area of 129 x 97 μm (760 x 572 pixels) with an axial step size of 76 nm, and giving an axial sensitivity of 20 nm with interpolation. To cover the full depth of the samples (approximately 20 μm), a single full field measurement of 760 x 572 pixels required 260 images, which took about 2 minutes to acquire and process. The uncertainty in measured surface height of the system varies typically from 20 to 300 nm depending on the structure measured, as well as the principal sources of error coming from the positioning uncertainty of the piezoelectric Z table, the camera noise and the influence of the algorithm.

Previous studies of the structure and chemical composition of the grown layers showed the formation of carbonate-containing HA on the modified and initial surfaces of the three materials [8,9,22]. LM revealed the formation of thick and heterogeneous layers consisting of white crystals (Fig. 4). The optical microscope was used before the measurements with stylus, AFM, SEM, TEM and CPM techniques in order to roughly estimate the layer thickness on all samples (modified and controls) and the calculated thickness was in the range of 14-25 μm.

SEM images (Fig. 5) showed the formation of heterogeneous layers. Higher magnification (inset) revealed the formation of clusters of sphere-like particles with an average diameter of 0.75 μm, and a random distribution of cavities throughout the layer volume.

Figure 5. Typical SEM image of the HA layer (magnification x500). Inset shows details (magnification x5000).

The HA topography was examined also with AFM by scanning area with the sizes of 20 x 20 μm (Fig. 6). The 2D, 3D images and surface profiles obtained showed that the grown layers were rough and irregularly covering the substrate surfaces. The layer peak-to-valley height R_t (total height between the peak of the highest point and the dip of the lowest valley) measured over the area of 20 x 20 μm was 948 nm. Surface profiling over a line yielded a value for the R_t of 600 nm (Fig. 6; the length of the profile line size shown in the 2D image is 28 μm). These values were considered as referring to the top layers since the tip could not reach the substrate surface due to the high layer thickness and roughness. Thus, it was not possible to measure the thickness of the grown layers using AFM. Nonetheless, additional information was obtained, revealing an average cluster diameter of 0.45 μm (Table 1), which was smaller than that obtained from the SEM images. Attempts to measure the layer thickness and roughness by stylus profilometry showed values for the thickness in the range of 1-5 μm, which were assigned to measurement of the top layers because the tip tends to dig into the layer.

TEM (Fig. 7) and XRD measurements revealed that the HA particles building the clusters contained nanosized (20-40 nm) needle-like HA crystals [9]. The layer thickness evaluated by TEM was in the order of 1-2 μm. The reason for this low value was that most likely the overgrown layers were detached from the initial layer during the sample preparation for the TEM imaging.

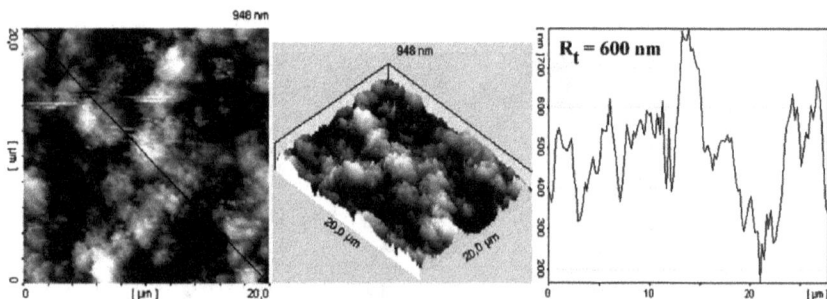

Figure 6. AFM results for the topography of the HA layers: 2D and 3D images (image size: 20x20 μm) show that the layers are rough and irregularly covering the substrate surfaces.

Table 3. HA particle diameters measured by AFM, SEM and CPM techniques.

	d average (µm)	d min (µm)	d max (µm)
AFM	0.45	0.2	0.9
SEM	0.75	0.4	1.1
CPM	1.9	1.0	3.3

These results displaying significant roughness and heterogeneity of the layers show why it was found to be difficult, destructive or time-consuming to measure the topography, thickness and/or profile of the grown HA layers by using stylus, AFM, TEM and SEM. Therefore, CPM was applied to obtain the necessary data and recorded representative images are shown in Fig. 8. The extended depth of field reflection image (Fig. 8 a) provides qualitative information about the homogeneity of the layers. The high contrast of the image was due to the difference in reflectivity between the substrate (white central area) and the HA particles (darker granulated areas). The variation in the contrast of the HA particles is due to the rounded aspect of the particles, the white dots corresponding to the upper flat part of the particles reflecting more light than the surrounding sloping parts. The grayscale height image (Fig. 8 b) obtained using the demodulation algorithm and filtering with a 5 x 5 median filter gave quantitative information for the layer. It confirmed that the layers were rough, quite heterogeneous, and varying from areas of bare substrate to partially covered substrate, and up to dense regions 14-20 µm in height. The filtered height image was used for the calculation of the 3D view (Fig. 8 c), which gave a qualitative impression on the layer topography. The measured surface line profile depicted in Fig. 8 d corresponding to the line in Fig. 8 b revealed the high layer thickness and roughness (the line length is 126 µm).

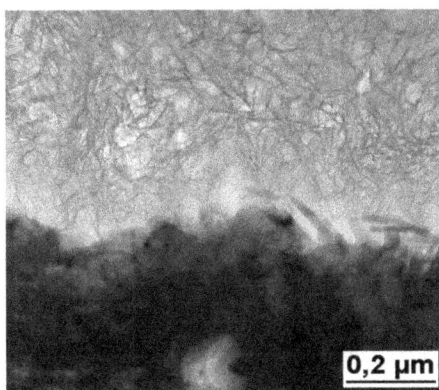

0,2 µm

Figure 7. TEM image of the HA layer (up) and the stainless steel substrate (down) shows the growth of needle-like crystals.

The layer formed on the thermally treated and control silicon samples showed comparable values (15.6 and 14.8 µm, respectively). In general, the stainless steel surfaces led to the formation of thicker layers in comparison to the silicon surfaces. The rms roughness showed comparable values for the modified and control stainless steel and silicon samples (4 µm), with the only exception being the thermally treated stainless steel sample (1.8 µm). Using the surface profiles we estimated the average diameter of the particles in the HA clusters to be between 1 and 2 µm (Table 3), slightly higher than the SEM results. The increasing measured particle size was detected in the order: AFM<SEM<CPM.

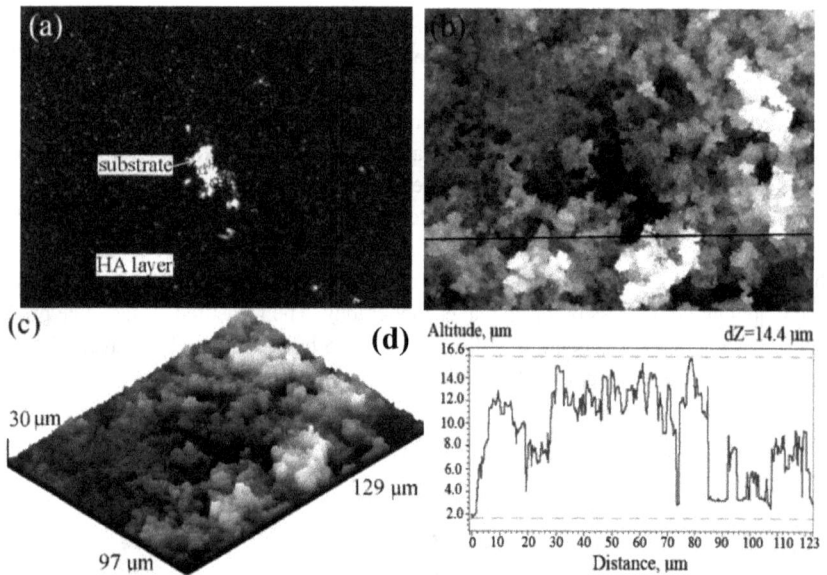

Figure 8. Representative images of the HA layers obtained using the CPM technique (magnification 50x; image size 129 x 97 µm) confirm that the layers are rough, quite heterogeneous, and varying from areas of bare substrate to partially covered substrate, and up to dense regions 14-20 µm in height. (a) Reflection mode – large depth of field image (qualitative view); (b) grayscale mode – filtered height image (quantitative view); (c) 3D mode (qualitative view); (d) surface profile corresponding to the line in (b) allows us to estimate the average particles diameter to be between 1 and 2 µm.

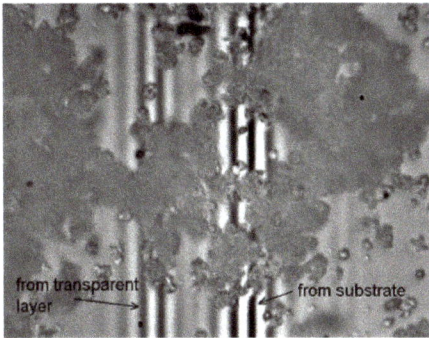

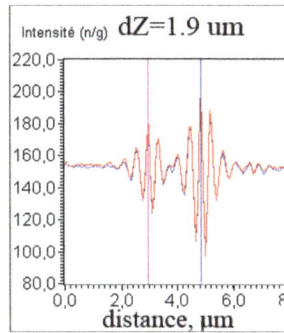

Figure 9. Double interference signal observed during scanning along the optical axis attributed to the presence of a transparent layer between the silicon substrates and the HA layer.

Figure 10. Double interference signal observed during scanning along the optical axis attributed to the presence of a transparent layer between the silicon substrates and the HA layer.

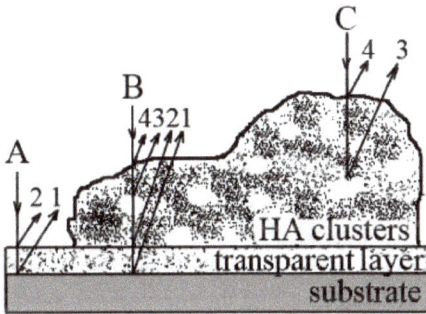

Figure 11. Model of the HA layer. From bottom to top: substrate, transparent layer, thin and thick HA cluster layers. The light beams reflected from the different interfaces are shown with numbers from 1 to 4.

When measuring the layers grown on the silicon samples, a second set of interference fringes was observed in certain regions as shown in Fig. 9. Scanning the sample along the optical axis (Z) and measuring the intensity at a single point (averaging over a 3 x 3 pixels window to reduce noise), a complex interference signal was observable at certain places (Fig. 10). The distance between the peaks – envelopes of the two signals reflected from the substrate-layer boundary and the layer-air boundary dZ gives the thickness of this layer - 1.9 µm. RBS identified the signal as silicon oxide layer. Such signals contain useful information concerning the structure and optical properties of the underlying material and interpretation depends on the complexity of the signal and the position within the image where the measurement is made. Using information from the height and the reflection images to choose the positions for making Z profiles, three different types of zones on the samples (A, B and C) have been identified and a model for the growth of the layer in the Z direction was drawn (Fig. 11). A transparent layer visible only under the interference microscope was present on the substrates (zone A). A thin layer of clusters (< 10 μm) consisting of HA sphere-like particles was formed on the transparent layer (zone B). Adherence of clusters to those already formed increased their total thickness to up to 20 μm (zone C). The fringe profiles associated with each zone contained fringe envelope peaks, which corresponded to different parts of the layers along the optical axis and are presented in Fig. 12:

- peak 2 from the 'transparent layer-air' surface in zone A or the transparent layer-thin HA cluster' interface in zone B. RBS was used to identify the transparent layer as being SiO_2 and its thickness was determined through simulations to be 1.3 μm. By using the CPM, the distance between the two peaks 1 and 2 was measured to be 1.9 μm which corresponds to an optical path difference $n.d$ (where d is the layer thickness and n the average refractive index of the transparent layer). These results agree with those from RBS, for a 1.3 μm thick layer of SiO_2 with a refractive index of just under 1.5;

- peak 3 from cavities buried within the HA clusters in zones B and C;

- peak 4 from the 'HA cluster-air' surface.

In areas where there was no cluster layer, the two peaks 1 and 2 of the Z profile and the smooth fringes in the XY image of these areas indicated the presence of the transparent layer corresponding to zone A. The second zone B in which the two peaks 1 and 2 could be observed together with a cavity (peak 3) and the 'HA cluster-air' surface corresponded to an area of thin cluster layer (< 10 μm). The slight shift of the two peaks 1 and 2 in zone B to the right compared with those in zone A indicated that the average refractive index of the thin cluster layer must be slightly greater than 1. If it was exactly 1, these peaks would be in the same place along the Z axis as those in zone A. If it was much more than 1, the peaks would be significantly shifted to the right. The third zone C, in which only the upper 'cluster-air' surface (peak 4) and a buried cavity (peak 3) could be observed corresponded to a thick cluster layer (> 10 μm). Peaks 1 and 2 corresponding to the buried transparent layer were not visible in zone C because of the lack of spatial coherence between the interfering signals due to the presence of the thicker cluster layer. White light speckle was observed in the XY image in zones B and C, indicating the presence of high surface roughness of the same order of magnitude as the wavelength of the light used (350 - 1100 nm).

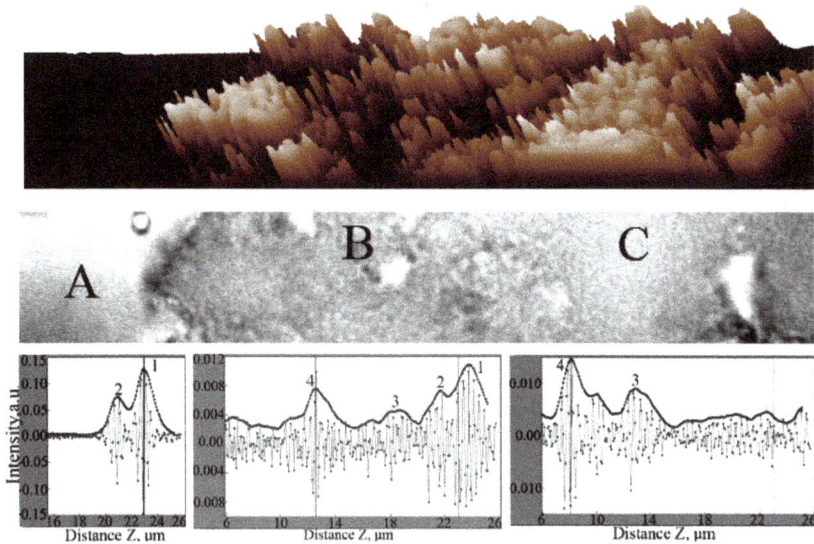

Figure 12. 3D and 2D images resulting from the Z scans along the depth of the HA layer with envelope peaks corresponding to: (1) the 'substrate- transparent layer' interface, (2) the 'transparent layer-air' surface or the 'transparent layer- HA cluster' interface, (3) a buried cavity, (4) the 'HA cluster-air' surface. The zones identified in peak 1 from the 'substrate- transparent layer' interface was observed in zones A and B. the XY image are: A, the 'substrate- transparent layer'; B, a thin layer of HA clusters (< 10 µm) and C, a thick layer of HA clusters (> 10 µm).

While CPM can be used successfully to measure the thickness and roughness of HA layers, one of the disadvantages of the technique is that it has a lower lateral resolution (0.45 µm) than AFM, SEM and TEM due to the effects of diffraction. When measuring HA clusters, one observes two effects: firstly, particles smaller than 0.45 µm are not imaged and secondly, particles slightly bigger than 0.45 µm in size are low pass filtered. The edges of the particles are thus rounded off and clusters tend to be merged in the resulting height image (Fig. 8 b), leading to a larger measured cluster size (1 µm to 2 µm) than that found in SEM (Fig. 5). A slight improvement in the lateral resolution of the measurement system in CPM would considerably improve the measurements. This could be achieved by using a higher numerical aperture objective (0.95) at shorter wavelengths (400 nm) for example, which would give a lateral resolution of better than 0.3 µm. In this case the CPM measurements of lateral particle size would approach those of SEM. An advantage of the results obtained by CPM is that much larger areas, 100 x 100 μm in a single measurement, are covered in comparison to those typically obtained by AFM. Even bigger areas can easily be covered using scanning and image stitching. Covering a larger area assures better statistical data of the surface roughness and lateral dimensions.

Difficulties in making measurements using CPM on these thick HA layers arise from the existence of different types of regions, giving classical interference fringes on smooth surfaces (Fig. 13 a), speckle-like fringes on rough surfaces and multiple fringe signals (Fig. 13 b) along the optical axis in the presence of buried layers. For example Fig. 13 shows the XZ images taken from the XYZ image stack. A lot of information is available in these kind of images concerning the HA layers. The different fringe envelopes along the Z axis indicate the positions of reflecting surfaces and interfaces. The position of the layer/substrate signal along the Z axis in relation to that of the substrate signal can provide information concerning the refractive index n. The presence of several fringe envelopes along the Z axis indicates the presence of cavities within the HA layer.

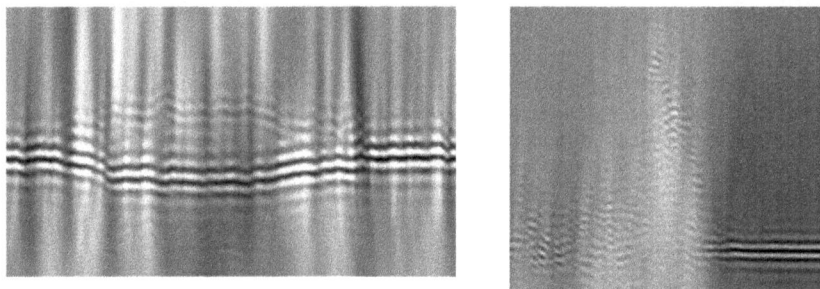

(a) double interference signal indicating presence of thin HA layer on the substrate.

(b) more complex interference signals indicating the presence of thicker layer of HA with cavities on the substrate.

Figure 13. XZ images of HA layers on stainless steel substrate using CPM.

The semi-transparency of HA, its high surface roughness and complex spongy aspect makes its structural characterization a challenge for any technique, even for WLSI micro-scopy where the interference fringes have low contrast because of the semi-translucent specificity of the material. In addition, there are several image degradation sources due to the camera. Basic post processing methods, such as image averaging, and dark and flat corrections, can be used to improve the quality of the fringe images by increasing the signal to noise ratio (SNR). Likewise, the image contrast can be enhanced with the hybrid High Dynamic Range (HDR) technique [25] by combining images with different exposures to obtain images with a greater depth range. Indeed, image processing techniques are already widely used in astronomy for the study of stars, galaxies and for the detection of exoplanets. The latter objects are very faint unresolved point sources that are initially lost in the image noise since they are so close to the star around which they orbit. However their orbits can be resolved, making individual exoplanets observable. In order to be able to measure the intensity of the planets, their position and their movements, complex procedures have been developed in the chain of image acquisition, post-processing and data reduction, thus allowing a significant reduction in the noise and an improvement in the resolution and power of detection of telescopes [26]. In the same way, we have begun the development and implementation of an analogous rigorous methodology in unlabelled optical nanoscopy, with the same aim of being able to observe separate unresolved structures, but in this case, concerning microscopic structures in materials with an FF-OCT system. This procedure has been called the "IMPROVE-Protocol" for IMage PRocessing Optimization for fringe Visibility Enhancement (IMPROVE), which consists in a successive application and combination of image processing techniques, as follows:

For the acquisition: (1) the high dynamic range technique is used with a combination of 10 different exposure times; (2) averaging of 20 frames is performed for each exposure time, and for the data reduction: (3) the dark correction is performed by subtracting from each image a dark image averaged 20 times, (4) the flat correction consists of the division of each image with a flat image averaged 20 times. For the data analysis, the individual fringe signals along Z, the raw 2D fringes (XZ image), the processed tomographic images (XZ image) and the heights in the 3D images are all available for studying the structure of the layer. After applying the developed IMPROVE-Protocol with an adapted Leitz-Linnik full-field optical coherence tomography system, normally unmeasured features were observed. The outcome of this procedure was a meaningful increase in the SNR of the images and an improvement in the power of detection of the optical system, enabling new sub-μm sized structures to be detected (improved image contrast) as discussed in more details in [27].

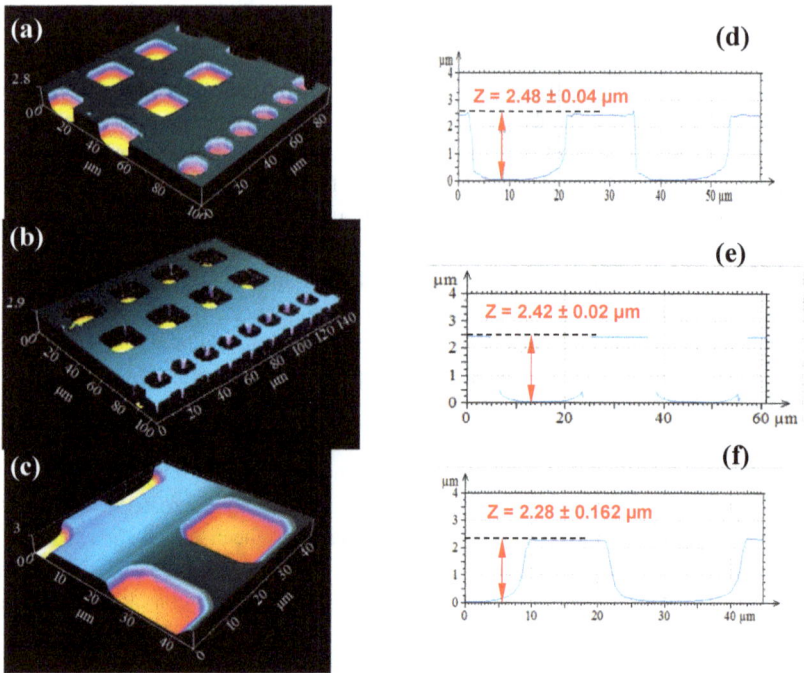

Figure 14. 3D images of the same etched squares in Si using (a) Leitz-Linnik microscope, (b) Zygo interferometer, (c) AFM microscope, as well as the corresponding line profiles (d-f) obtained from the 2D images

Further improvement of the CPM technique allowed us to obtain more precise information concerning the HA layer when compared to the AFM and classical interferometer. To be able to measure the same zone with the different techniques, a silicon wafer etched with a square pattern to an average depth of 6.6 µm was used as a substrate on which to grow HA, with enough layer to be measured while leaving spaces to be able to recognize the pattern. The etched silicon samples were left in the SBF for 3 hours and 25 min at 37 °C ad pH 7.4 under continuous stirring. The CPM interference microscopy system used was developed at ICube, University of Strasbourg. The system based on a Leitz-Linnik microscope equipped with a Linnik type objective (x50, NA = 0.85) gives a lateral resolution of $R_{lat} = 0.43$ µm in visible light, an axial resolution of R_{ax} = 1 to 15 nm depending on the algorithm used and nature of the surface to be measured and a maximum field of view of 180 x 140 µm with a Prosilica CE1380 camera (1360x1024 pixels). The software used (CPM 2.2) for the control, acquisition and processing was developed in-house using LabView. The algorithm used for fringe processing and surface roughness characterization was an improved version of the visibility measurement technique [17] derived from Teager-Kaiser energy operators [18]. Rapid image acquisition was used (sequence of successive images) with axial steps of 90 nm, slight noise reduction (low pass 3x3 Gaussian filter) and envelope peak interpolation using second order spline fitting. Zygo interference microscope's field of view is 140 x 110 μm and the system has a quoted axial resolution of ~ 0.1 nm with a precision of < 0.75 % over 150 μm. The field size obtained with the AFM was 45x45 μm for 512x512 pixels, giving a lateral resolution of 90 nm. This value can be improved by scanning over a smaller area with a higher number of pixels. For example, scanning over 2x2 μm for 1000x1000 pixels gives a lateral resolution of 2 nm. For analyzing and comparing the measurements from the different microscopes, MountainsMap® 6 software (from Digital Surf) was used. This enables false color images of the height data, 3D views, 2D surface line profiles and height and roughness measurements to be made in a uniform way from the different measurement sources. For comparing the line profiles from the same place, a line profile 45 μm in length, corresponding to the width of the AFM measurements, was taken from the middle of the same etched square in the measurements from each of the microscopes. The square was identified using the numbering system.

The 3D measurements on the same area of bare patterned Si wafer with the corresponding line profiles obtained from the 2D images (Fig. 14) show broad similarities in the results. As would be expected, square depressions of a similar size were found in each case while on closer study, several differences can be observed. The most significant difference is the variation between the values of the average depths of the etched squares found by teach measurement technique. The average measured depth with Leitz-Linnik, Zygo interferometer and AFM was 2.48±0.04 μm, 2.42±0.02μm and 2.28±0.16 μm respectively. The estimated percentage uncertainty was ± 1.6 %, ± 0.75 % and ± 7 %. The estimations of the uncertainties of measurements for each technique were made in the following way. For the Leitz-Linnik measurements, the main contributions to the uncertainty come from errors in the piezo step positioning, the envelope determination algorithm and the reference mirror flatness, resulting in an overall uncertainty of ± 0.04 μm (± 1.6 %). Concerning the Zygo interferometer, the uncertainty of ± 0.02 μm is calculated from the quoted accuracy of ± 0.75 % of the depth measurement. For the AFM measurement the main contributions to the measurement uncertainty come from the non-linearity at depths greater than a few hundred nm together with noise from acoustic vibrations, leading to an estimated uncertainty of ± 0.16 μm (± 7 %). The uncertainty of the AFM measurements could be improved by optimizing the choice of the control parameters for the measurement and for that purpose the AFM microscope has been placed in an acoustic enclosure, which will also improve the uncertainty values.

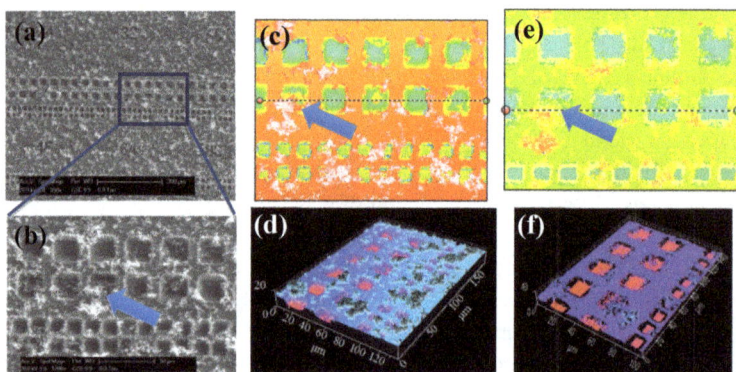

Figure 15. Comparison of the surface shape of the same place of HA layer deposited on etched squares in Si measured using (a,b) SEM, (c,d) Leitz-Linnik microscope and (e,f) Zygo interferometer; blue arrows indicate same cluster of HA.

Another significant difference between the 2D profiles can be observed near to the measurements of the edges in all three cases. For both of the interferometric measurements, edge effects are visible. For the Leitz-Linnik this error is in the form of an over estimation of the top of the edge (see line profile in Fig. 14 d) and for the Zygo it is in the form of an under estimation at the bottom of the edge (line profile in Fig. 14 e). These errors appear to be similar to the well-known "batwing" artifacts when measuring a step height near to the coherence length of the light used due to mixing of signals coming from the top and bottom of the edge [20]. In addition, for the Zygo results the missing measurement points between the top and bottom of the edges are due to the lack of fringe signals on the steep slopes because of the limited numerical aperture of the objective. Finally, for the AFM measurements (see line profile in Figure 14 f), the edges can be seen to be rounded off which is a result of the convolution between the tip shape and the edge of the etched square. The degree of rounding off was also found to vary for different tip scanning speeds and sample orientations.

For the HA layers deposited on etched silicon patterns, measurements were not possible using AFM because of the high layer thickness and depth of the etched squares (6.6 µm) and the limited dynamic range of 7 μm of the instrument. It is well known that HA is very difficult to characterize with AFM [22]. A first comparison of the different 3D measurements on the same region (Fig. 15) again shows certain broad similarities in the results, indicating a partially rough layer on square depressions. The presence of individually recognizable large clumps of HA, such as the one indicated by the large arrow in Fig. 15, confirms that exactly the same area had indeed been measured by each technique and the success of the numbered multi-scale square pattern used.

A comparison of the 2D line profiles made at the same place (Fig. 16) between the measurements on the Leitz-Linnik and Zygo also shows broad similarities as well as the presence of differences and artefacts. An exact comparison of the height measurements of the HA layer between the two systems is more difficult than just with the bare etched silicon squares due to the greater variation in surface height with axial position of the rough HA layer. Nonetheless, by careful positioning of the cursor lines in the two measurements, the two profiles obtained (Fig. 16) can be seen to be fairly similar. A comparison of the heights at the places indicated revealed that the average measured depth with Leitz-Linnik and Zygo interferometers was 5.05 ± 0.13 μm and 5.06 ± 0.12 μm, respectively. The estimated percentage uncertainty was $\pm$ 4% for both techniques. The uncertainties in measurement were calculated as for the bare Si wafers. The higher values of uncertainty in measurement on the HA layer compared with the results on the bare silicon are due to the inaccuracy from the placing of the measurement cursor in the image and the large lateral variation in roughness of the HA.

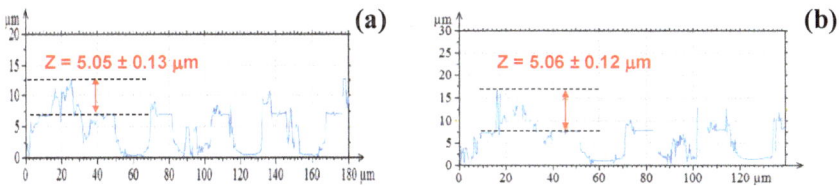

Figure 16. Comparison of 2D profiles of HA layer deposited on etched squares in Si measured using (a) the Leitz-Linnik microscope and (b) the Zygo microscope.

Another difficulty of measuring layer growth is the physical growth conditions. Most of the previously mentioned techniques (SEM, AFM, stylus, TEM, etc.) are generally used in dry, *ex situ* conditions. *In situ* measurement is more complicated by the limited space available due to a growth chamber and the difficulty to approach the sample surface. In the case of growth in a liquid such as the SBF, the presence of the growth solution also limits the type of optics that can be used. A new optical measurement head is being developed that is suitable for use in liquid immersion conditions, with the view of measuring layer growth or modification in biomaterials. In order to be able to measure the growth of biomimetic HA layers *in situ*, an immersion head is being specifically developed, shown in Fig. 17. A pair of immersion objectives is used in a Linnik configuration. The sample observation objective will be placed in a protective head containing distilled water and a window that will be immersed in the SBF solution. In this way, the observation window can be cleaned or replaced between growth experiments. The head is under development at ICube, University of Strasbourg, France.

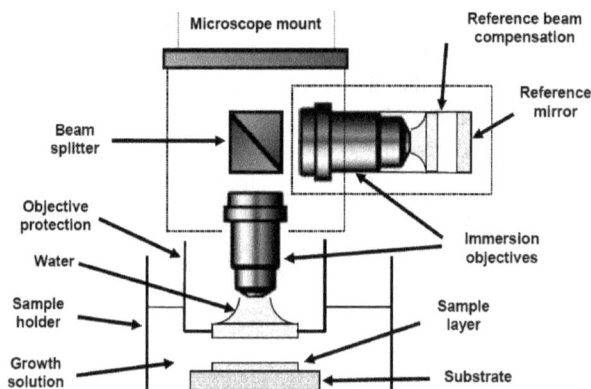

Figure 17. Adapted immersion Linnik interferometer head for characterizing in situ layer growth in liquid conditions is being under development.

3.   Summary

This work has shown that an efficient CPM system can be built for measuring topography, surface profiles and roughness of thick biomimetic HA layers (height above 10 µm), based on WLSI, combined with precise height stepping of the sample, and image processing. CPM provides rapid, non-destructive measurements and does not need any sample preparation. Various algorithms for image stacking and processing have been used for obtaining optimum results. HA investigated in this work by CPM presents a challenge for this technique because of its semi-translucence, high roughness and the presence of cavities, which in certain configurations affects the axial precision.

Nevertheless, CPM can be efficiently used to obtain complementary qualitative and quantitative information on thick and rough HA layers together with standard imaging techniques (i.e. SEM, AFM, TEM, LM). In addition, better statistical data can be obtained by CPM due to the large measured area (100 x 100 μm or higher). The latest results for optimizing the measurement conditions in order to reduce different sources of errors and to begin to exploit the additional information in the Z profiles due to different structures present within the layer have been presented in this chapter. Work is in progress to further improve the lateral resolution and to extend the complex structure identification from single point axial measurements to full-segmented image analysis using improved algorithms. Moreover, the combination of the real time system and immersion head will enable the study of the growth of HA layers *in situ*.

References

[1]   Handbook of analytical methods for materials, Materials evaluation and engineering Inc., 2001, MN, USA

[2]   F.H. Jones, Teeth and bones: applications of surface science to dental materials and related biomaterials, Surf. Sci. Reports 42 (2001) 75- 93 https://doi.org/10.1016/S0167-5729(00)00011-X

[3]   D.B. Williams, C.B. Carter, Transmission electron microscopy, Plenum, 1996. https://doi.org/10.1007/978-1-4757-2519-3

[4]   T. Thompson, The Fundamentals of Raman Spectroscopy, Laser Focus World 224 (2012) 1-15

[5]   E. Bright Wilson, J.C. Decius, P.C. Cross, Molecular Vibrations, McGraw-Hill Book Company, 1955.

[6]   Basics of X ray Diffraction, in: X-ray diffractometer, ch. 7, Wordpress.com, https://mkmcatalysis.files.wordpress.com/2014/09/xrd-sajid.pdf (last accessed November 2016)

[7]   T. Kokubo, H. Kushitani, S. Sakka, T. Kitsugi, T. Yamamuro, Solutions able to reproduce in vivo surface-structure changes in bioactive glass-ceramic A-W, J. Biomed. Mater. Res. 24 (1990) 721-734. https://doi.org/10.1002/jbm.820240607

[8]   L. Pramatarova, E. Pecheva, V. Krastev, F. Riesz, Ion implantation modified stainless steel as a substrate for hydroxyapatite deposition. Part I. Surface modification and characterization, J. Mater. Sci.: Mater. Med. 18 (2007) 435-440 https://doi.org/10.1007/s10856-007-2002-4

[9]   L. Pramatarova, E. Pecheva, V. Krastev, Ion implantation modified stainless steel as a substrate for hydroxyapatite deposition. Part II. Biomimetic layer growth and characterization, J. Mater. Sci.: Mater. Med. 18 (3) (2007) 441-447 https://doi.org/10.1007/s10856-007-2003-3

[10]   P.C. Montgomery, in: Near field optics and nanoscopy; J.P. Fillard (Ed.), World Scientific, Singapore, 1996, ch. 4.5, pp. 197-214.

[11]   P. Caber, Interferometric profiler for rough surfaces, Appl. Optics 32 (1993) 3438-3441. https://doi.org/10.1364/AO.32.003438

[12]   P. De Groot, L.J. Deck, Surface Profiling by Analysis of White-light Interferograms in the Spatial Frequency Domain, J. Modern Optics 42 (1995) 389-401. https://doi.org/10.1080/09500349514550341

[13]   K.J. Larkin, Efficient nonlinear algorithm for envelope detection in white light interferometry, Opt. Soc. Am. A 13 (1996) 832-843. https://doi.org/10.1364/JOSAA.13.000832

[14]   P.C. Montgomery, J.P. Ponpon, M. Sieskind, C. Draman, Surface morphology analysis of HgI2 and PbI2 with interference microscopy – the challenges of measuring fragile materials and deep surface roughness, phys. stat. sol. C 0 (2003) 1044-1050. https://doi.org/10.1002/pssc.200306236

[15]   P.C. Montgomery, J.P. Fillard, Peak fringe scanning microscopy (PFSM): submicron 3D measurement of semiconductor components, Proc. SPIE 1755 (1992) 12-23. https://doi.org/10.1117/12.140772

[16]   M.B. Sinclair, M.P. de Boer, A.D. Corwin, Long-working-distance incoherent-light interference microscope, Appl. Optic 44 (2005) 7714-7721. https://doi.org/10.1364/AO.44.007714

[17]   M. Roy, C. Sheppard, P. Hariharan, Optics Express 12 (2004) 2512-2516. https://doi.org/10.1364/OPEX.12.002512

[18]   A. Dubois, L. Vabre, A.C. Boccara, E. Beaurepaire, Appl. Optics 41 (2002) 805-812 https://doi.org/10.1364/AO.41.000805

[19]   A. Dubois, K. Grieve, G. Moneron, R. Lecaque, L. Vabre, A. Boccara, Appl. Optics 43 (2004) 2874-2883. https://doi.org/10.1364/AO.43.002874

[20]   L. Vabre, V. Loriette, A. Dubois, J. Moreau, A. Boccara, Opt. Lett. 27 (2002) 1899-1901. https://doi.org/10.1364/OL.27.001899

[21] A. Benatmane, P.C. Montgomery, E. Fogarassy, D. Zahorski, Interference microscopy for nanometric surface microstructure analysis in excimer laser processing of silicon for flat panel displays, Appl. Surf. Sci. 208-209 (2003) 189-193. https://doi.org/10.1016/S0169-4332(02)01367-3

[22] E. Pecheva, P. Montgomery, D. Montaner, L. Pramatarova, White light scanning interferometry adapted for large area optical analysis of thick and rough hydroxyapatite layers, Langmuir 23 (2007) 3912-3918 https://doi.org/10.1021/la061593f

[23] A. Leong-Hoi, P.C. Montgomery, B. Serio, P. Twardowski, W. Uhring, High-dynamic-range microscope imaging based on exposure bracketing in full-field optical coherence tomography, Opt. Lett. 41 (2016) 1313-1316 https://doi.org/10.1364/OL.41.001313

[24] R. Galicher, C. Marois, B. Macintosh, T. Barman, Q. Konopacky, M-band imaging of the hr 8799 planetary system using an innovative loci-based background subtraction technique, Astrophys. J. Lett. 739 (2011) L41 https://doi.org/10.1088/2041-8205/739/2/L41

[25] P.C. Montgomery, F. Salzenstein, D. Montaner, B. Serio, P. Pfeiffer, Optical Measurement Systems for Industrial Inspection, Proc. SPIE 8788 (2013) 87883 G11 https://doi.org/10.1117/12.2020560

[26] F. Salzenstein, P.C. Montgomery, D. Montaner, A. Boudraa, Teager-Kaiser Energy and Higher-Order Operators in White-Light Interference Microscopy for Surface Shape Measurement, Eurasip. J. Appl. Signal Process. 17 (2005) 2804 https://doi.org/10.1155/ASP.2005.2804

[27] A. Leong-Hoi, P.C. Montgomery, B. Serio, W. Uhring, E. Pecheva, Improvement in measurements of hydroxyapatite layers by hybrid high dynamic range image processing in white-light interferometry, Materials Today Proceedings, 2017, in press

CHAPTER 3

Laser-Liquid-Solid Interaction Method: Stimulation of the Growth of Biocompatible Layers by Using External Laser Energy

Abstract

Important in studying the biomineralization and the formation of biomaterials such as HA is the investigation of the surface and its appropriate structural and/or chemical modification, leading to desired bioreactions and improvement of the biomaterial's performance since it is intended to be placed in various (aggressive) liquid media. The traditionally applied method for the growth of biomimetic HA is through the materials immersion in a supersaturated SBF solution in order to mimic the process of biological apatite formation [1]. Despite its similarity to the mineral formation in nature, the method requires a long time to obtain CaP precipitates. The application of external fields have been used to influence the HA formation *in vitro* (ultrasound, ultraviolet and microwave irradiations [2–4]). On the other hand, over the last decades, laser sources of energy have proven to be major tools for surface processing of a large variety of materials, from metals to ceramic and polymers. Pulsed lasers have proven to be an invaluable tool in research and development, and even in the discovery of new thin film materials because of the unique nature of the laser-material interaction (e.g. power, energy, time, special profiles, laser repetition rate, etc.). The feasibility of using laser ablation in transparent liquids for nanostructured materials fabrication has been demonstrated [5-7]. Laser sources of energy have been efficiently used for processing living tissues, synthesis of biocompatible materials, biofilm modification, bone defect healing, modification of root canal dentine, laser sterilization of dental implants, etc. [8-13]. Lasers have also been employed to increase the long-time performance of Ti dental implants through assuring cleanness, specific microrelief and a stable oxide layer on the surfaces [14].

The task of the work described here was to develop a method for laser enhanced deposition in an aqueous environment in order to obtain bioactive nanostructured HA resembling the biological apatite. The idea for a synergistic effect of the simultaneous exposure of a solid surface to radiation and water was embraced and a laser was chosen due to its possibility for precise, and flexible irradiation of small, and complex shapes and for its high power density [15,16]. Utilizing the LLSI method, hierarchically organized surfaces on both micro- and nanoscale that are typical for nature can be modelled.

Keywords

Laser Interaction with Materials, External Energy, Influence on Biomaterials Growth, Optical Emission Spectroscopy

Contents

1.   Principle of the Method

The experimental set-up included an open deposition system where the solution was placed in a reaction container and the solution temperature was controlled by a thermostat (Fig. 17). The substrates were immersed one by one on the bottom of the container with 50 ml of the solution (layer thickness above the samples was 20 mm) at room temperature and simultaneously irradiated by a laser. CuBr pulsed vapor laser (Fig. 18; λ = 578.2 nm, laser energy per pulse 1.2 kW, output power 330 mW, laser power density 50 MW/cm^2, pulse duration of 30 ns, pulse repetition rate 19 kHz, laser beam diameter in the focal spot 50 μm), equipped with a precise scanner (velocity of 1000 mm/s) was used in the experiment. The laser beam was directed perpendicularly to and focused on the substrate surface through the solution. By using the scanning system, a surface pattern of seven concentric squares, each separated by a distance of 200 μm was formed at the edges of each sample as seen in the upper left corner of Fig. 17, and the middle of the samples was not irradiated. The chosen design was repeated 100, 200 or 500 times in order to study the influence of the duration of laser irradiation on the layer formation after the samples soaking in the SBF. The interaction time of the laser beam with each sample was less than 5 min. One group of samples was taken out of the SBF immediately after the end of the LLSI in order to study the instantaneous effect of the laser irradiation. Another group of samples was subsequently immersed in 500 ml total volume of fresh SBF for up to 24 hours under physiological conditions (37 °C, pH 7.4). A group of non-irradiated samples was put in a separate container with SBF (37 °C, pH 7.4) and used to grow control CaP layers for various time durations comparable to these of the laser-irradiated samples using the biomimetic approach. Before characterization all groups of samples were washed with distilled water and dried in air. The laser applied in our work is technologically available and well established, and it has been chosen also due to its possibility for precise and flexible irradiation of small and complex shapes peculiar to many implants, as well as for its high power density.

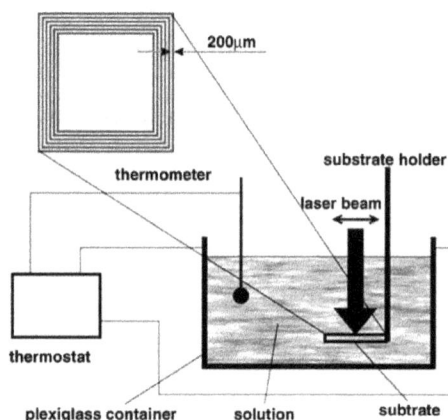

Figure 17. The set-up used in the LLSI process and design of the laser irradiation (upper left corner).

Samples from stainless steel AISI 316, n-type silicon with (100) orientation and silica glass Herasil subjected to standard mechanical treatment [17] were obtained for laser irradiation. To prepare the SBF two base solutions were prepared by using reagent-grade chemicals. The first one consisted of NaCl (15.99 g/l), KCl (0.45 g/l), $CaCl_2-2H_2O$ (0.74 g/l) and $MgCl_2-6H_2O$ (0.61 g/l) dissolved in 1 l of distilled water. The second one included $Na_2SO_4-10H_2O$ (0.32 g/l), $NaHCO_3$ (0.71 g/l) and $K_2HPO_4-3H_2O$ (0.69 g/l) dissolved in 1 l of distilled water. The pH of the two solutions was buffered at pH 7.4 with trishydroxymethylaminomethane buffer or hydrochloric acid [1]. Separately stored, the two solutions did not yield precipitation. Before immersion of the samples, equal quantities of the two base solutions were mixed to give the final SBF solution. It has to be noted here that NaCl constitutes 80% of the salts dissolved in the precursor solution.

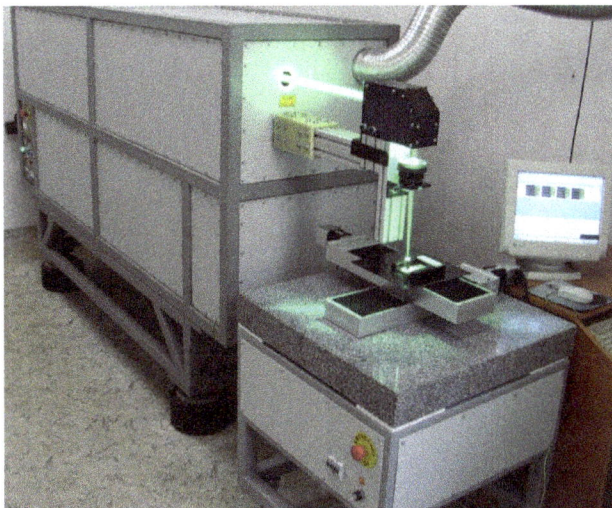

Figure 18. CuBr laser used in the LLSI experiments, produced by Pulslight Ltd.
(www.pulslight.com)

SEM (JSM-25 SIII) and optical reflection microscope (Olympus BX40) were used to characterize the layers grown on all materials after the LLSI process and after the subsequent soaking in the SBF. The elemental composition of the layers was measured by EDX (JEOL JSM- 840). The layer structure was studied by Raman spectroscopy (Jobin Yvon Horiba microspectrometer; $\lambda=532.14$ nm) and XRD (Siemens D5000, Bruker GmbH, Germany, $\lambda=1.54A$ °, $2\theta=20$–60°, step size of $2\theta=0.1$°; JCPDS file No 09–432). Absorption of SBF was measured with µ-Quant spectrometer (range of 200–1000 nm). Ca and P concentrations in the SBF were investigated by inductively coupled plasma mass spectrometry and ion chromatography. Roughness and height profiles were estimated by CPM (Leitz microscope Wetzlar with Linnik objective of 50x, $\lambda = 350$–1100 nm, dynamic range 100 mm, axial and lateral resolution of 10 nm and 0.45 m, respectively). The elasticity and hardness of the CaP layers was measured by a scanning force microscope (SFM, NanoScan, Russia, loads 200-500 µN).

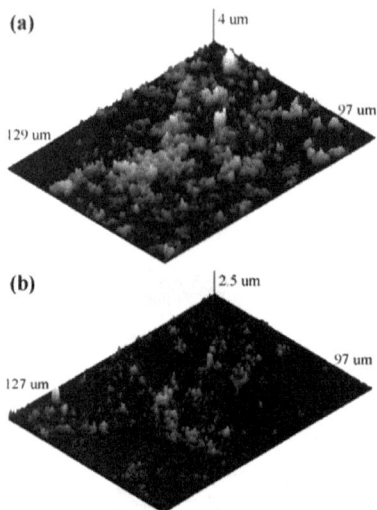

Figure 19. CPM images of the grown for 6 hours CaP layers on: (a) LLSI-treated and (b) control glass samples.

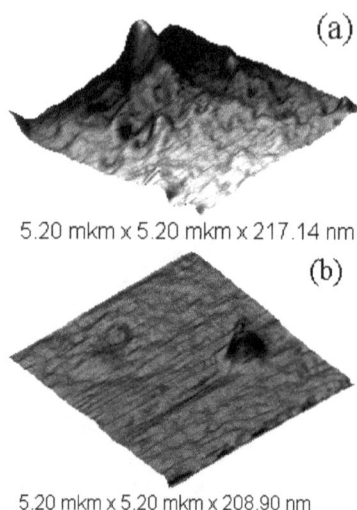

Figure 20. Nanoscan images of the (a) LLSI-treated and (b) control samples with CaP grown for 6 hours.

An inhomogeneous layer was grown on the glass surfaces from the SBF, either by applying the LLSI (Fig. 19 a) or without it on control samples (Fig. 19 b) as shown by the CPM images. The average layer thickness and rms roughness on the laser-treated samples were 3.7 and 0.9 µm, respectively, and on the non-treated samples these values were 2.3 and 0.5 µm. Laser-treated surfaces yielded higher CaP roughness. Nanoscan images of the layers on laser-irradiated and non-irradiated surfaces are shown in Fig. 20. The SFM technique showed that upon increasing the load from 200-500 µN, the elasticity of the laser-irradiated samples increased from 34-119 GPa, and for non-irradiated samples it was 37-56 GPa. The hardness of the samples also increased: under the maximum applied load of 500 µN, the measured value was 7.2 for the laser-treated and 3.6 GPa for the non-treated samples.

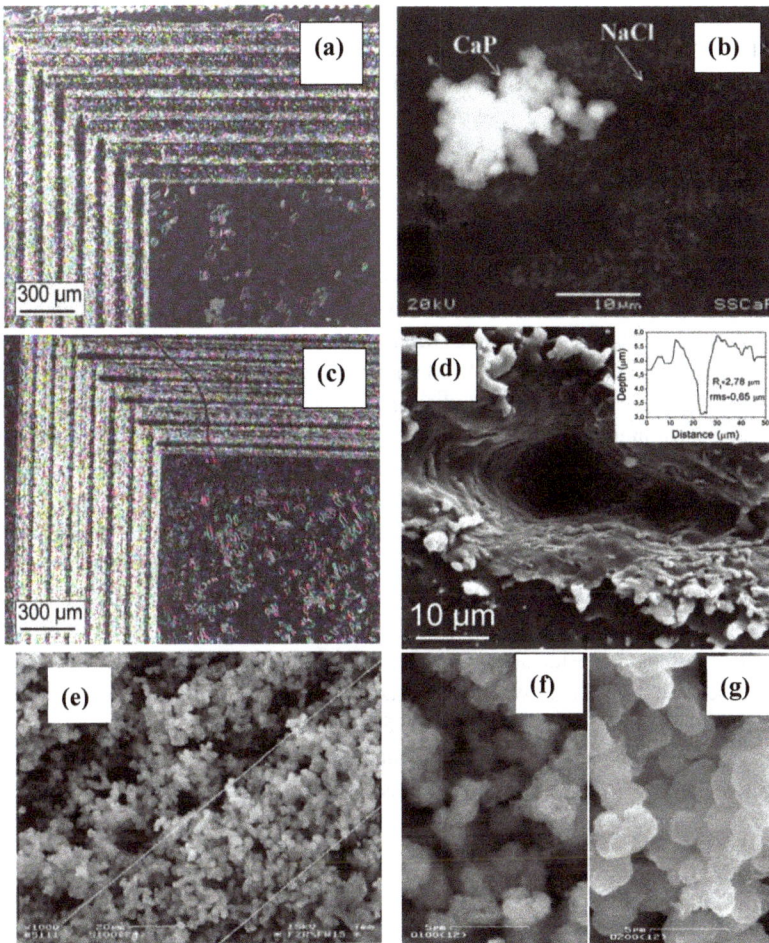

Figure 21. (a) LM image of a Si surface immediately after the LLSI process; (b) detail of the surface in (a) obtained by SEM image; (c) LM image of the laser irradiated area of the Si sample subsequently immersed in the SBF for 2 hours; (d) detail within the laser stripe obtained by higher magnification with SEM; (e) SEM image of a Si sample soaked in SBF after the end of the LLSI for 12 or 24 hours, scale bar 20 μm; (f, g) network of white particles deposited after 100 (f) and 200 repetitions (g) of the chosen laser design on glass, scale bar 5 μm.

LM and SEM imaging added additional details to the CPM and Nanoscan results (Fig. 21). LM (Fig. 21 a) depicts the laser lines on the Si surface and a predominant formation of white particles in the irradiated area. SEM images showed the formation of tiny white dots, as well as bigger white particles grouped in clusters (Fig. 21 b). EDX spectroscopy revealed their chemical composition: the tiny dots were composed of NaCl (higher magnification revealed they had the shape of dendrites) and the clusters consisted of Ca and P [18]. It was hypothesized that both type of particles observed in Fig. 21 b were seeds or nuclei formed due to the applied LLSI, which further facilitated the growth of the CaP layer when the samples were subsequently immersed in the SBF: LM image showed that the number of nuclei increased even for only 2 hours in the precipitated solution which means that the material has been activated by the laser irradiation (Fig. 21 c). Higher magnification obtained with SEM within the irradiated stripes revealed a place where the laser has been interacting with the surface: the high laser energy focused at the surface caused local melting of material thereby inducing surface roughness (Fig. 21 d; the same images were observed for the stainless steel samples). With 24 hours of soaking after the end of the LLSI, a white inhomogeneous layer covered the whole surface of the samples (Fig 21 e). SEM image revealed sphere-like particles grouped in clusters and forming a porous network with an average diameter of the particles of about 1.5 μm independently on the duration of soaking (12 or 24 hours). Comparison with the layers grown for the same duration by the biomimetic process showed higher diameter of 3-5 μm. The laser irradiation considerably stimulates the deposition process. The evidence for this can be seen in Fig. 21 e, where the irradiated stripe is outlined and its width corresponds to the diameter of the laser spot (50 μm). Denser and more homogeneous layer consisting of white particles grouped in clusters was observed within the irradiated stripe while the layer was less homogeneous outside the stripe. The repetition of the laser design resulted in accelerated layer growth as shown for the glass surface where bigger clusters were formed after more repetitions (Figs. 21 f, g).

In the SBF solution NaCl salt constitutes 80% of the salts as pointed out in the beginning of this chapter. After dissolving all salts in distilled water, HCl was used to adjust the solution pH. Adding HCl to an aqueous solution in which NaCl salt is dissolved according to reaction (1):

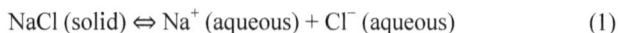

$$NaCl\ (solid) \Leftrightarrow Na^+\ (aqueous) + Cl^-\ (aqueous) \qquad (1)$$

drives the equilibrium of this reaction back, and as a result solid NaCl crystallizes [19]. Probably stable CaP nuclei formed due to the applied laser energy and they allowed the enhanced formation of the white CaP particles within the laser irradiation time (5 min). These particles further facilitate the growth of a thicker CaP layer when the laser-irradiated samples were left in the SBF for up to 24 hours. Such fast CaP formation was not observed when the samples were simply soaked in the SBF for the same time [17]. Regarding the pulsed laser irradiation of the materials, there are many experiments described in the literature in which laser irradiation has been utilized in a solution of soluble precursors, with different wavelengths, with or without a substrate, and the aim has been to form nanoparticles or to grow nanoparticle layers [20-24]. In our experiments, we obtained enhanced CaP formation on various types of surfaces as a result of the laser irradiation.

Up to now, a very limited number of works concerning 'laser–liquid–solid' interactions were published [25-30]. In the LLSI method utilized in our work, a complex interaction of the type 'laser–liquid–solid substrate' occurs. The surface of the materials after the LLSI process was observed by SEM and the image of the Si surface is shown in Fig. 21 d. After the multi-pulse processing with the focused CuBr laser beam (power density was 60 MW/cm^2), the surface was locally melted and ablated, thus yielding the formation of a crater with corona structure at its edges. As estimated by CPM (inset profile in Fig. 21 d), the depth of the crater was 2.78 μm and the rms roughness measured over a line crossing the crater transversely was 0.65 μm. The shape of the crater was typical for a Gaussian beam profile in the focal point, as those used in our experiment. The same SEM images of crater formation were observed for the stainless steel surface but the degree of melting and ablation was weaker. It is common in materials processing by high intensity nanosecond laser pulses that ablation occurs via extensive melting and evaporation [25-30]. In the applied power density range, plasma was formed and the molten material layer was pushed around the crater as a radial hydrodynamic flow due to the high recoil plasma pressures. Generally, mechanical irregularities introduced on the material surface are an advantage in the crystal growth processes because nuclei are easily attached to sites where surface defects exist [31]. This can explain the predominant attachment of CaP particles observed after the LLSI and revealed in Figs. 21 a and b. Another advantage of the crater formation was the enlargement of the effective interfacial area of the materials. Mechanical roughness and high surface area are known to play a significant role in stable layer formation, in anchoring cells and connecting together surrounding tissues, thereby leading to a shorter bone healing period. However, as a result of the ablation, rims were formed at the crater edge, which is an inconvenient effect because the rims may break off the material surface during an implanting procedure and contaminate the surrounding biological tissues. Observation of the surface chemistry changes caused by the laser irradiation was a subject of the work described in section 2 of this chapter. Nonetheless, a decrease of the surface carbon contamination due to a cleaning effect of the laser beam, as well as surface oxidation was expected and they are both beneficial in facilitation the HA formation.

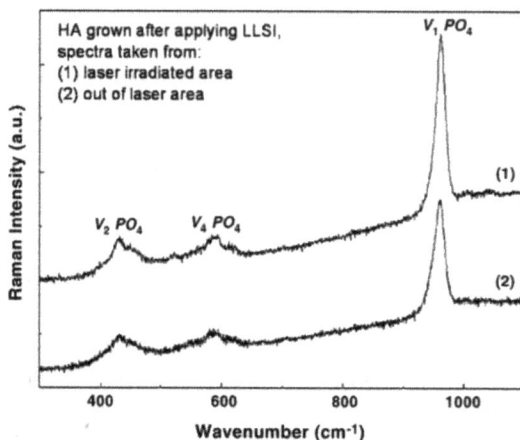

Figure 22. Raman spectra of laser-irradiated (1) and non-irradiated (2) samples showed the formation of a CaP layer and an increased layer thickness after the LLSI.

On one hand, there was the aqueous SBF solution, participating in the LLSI process. It is known that water is transparent for the visible light such as the one of the CuBr laser used in the experiments (578.2 nm). Absorption of the laser energy in the solution could appear because of the dissolved salts needed for the SBF preparation and the water. The light absorption of freshly prepared SBF and of the two base solutions used in the SBF preparation was measured in the range of 200–1000 nm and no absorption peaks were observed. Measured Ca and P concentrations of freshly prepared SBF showed a considerable decrease after the laser irradiation: fresh SBF had 136 mg/l Ca and 235 mg/l P and after the irradiation the concentrations decreased to 68 mg/l and 82 mg/l, respectively. It is known that when a precipitation begins in a supersaturated solution the ion concentrations decrease [31]. Thus, it was confirmed that the laser irradiation in the LLSI process resulted in the formation of CaP particles. The higher decrease of the P concentration indicated that it was preferentially consumed which can explain the low Ca:P ratio obtained by EDX (Ca:P < 1.0). The Raman spectra of the laser- irradiated (Fig. 22, spectrum 1) and non-irradiated samples (spectrum 2) were recorded from a laser stripe and in non-irradiated area. In both areas, they showed the formation of a CaP thin layer through the P-O vibrational modes of the PO_4 group. The stronger peak at 961 cm^{-1} was attributed to the v_1 P–O symmetrical stretching in PO_4^{3-}, and lower intensity peaks at

430, 590 and 1090 cm^{-1} were due to v_2, v_4 and v_3 P–O vibrational modes. Higher layer thickness was observed from the signal recorded in the laser stripe as seen from the higher intensity of the main P-O peak at 961 cm^{-1}. This result was assigned to the stimulation effect that the laser irradiation caused in the process of CaP nucleation and growth in the irradiated areas.

FTIR spectroscopy also revealed the formation of CaP (Fig. 23) through the utilization of the two methods: biomimetic approach (sample soaking in the SBF) and LLSI. The FTIR reflection spectra (Figs. 23 a, b) recorded at different time intervals demonstrate the kinetics of layer growth on the steel samples in the region 800 - 4000 cm^{-1}. The spectra in Fig 5a were recorded after sample horizontal soaking (hs) in the SBF for periods of 4 or 24 hours. The v_1 and v_3 P-O stretching vibrations characteristic of HA were registered in the region of 963 - 1113 cm^{-1}. Incorporation of carbonate (CO_3) after 4 hours of soaking was deduced from a shoulder at 874cm^{-1}, characteristic of v C-O bending in CO_3. It is possible that an acid CaP phase (i.e. HPO_4-containing) was also present in small amounts in the CaP layers but since the wavenumber of the P-OH mode coincides with the v_2 C-O mode, this could not be clearly concluded. With increasing time of soaking (i.e. after 24 hours), a weak peak splitting at 1422/1454 cm^{-1}, characteristic of v_3 C-O stretching in biological apatite was detected. A band at 1650 cm^{-1} and a broad band from 2500 to 3700 cm^{-1} in all spectra contributed to the H-O-H vibrations in adsorbed H_2O. It is known that biological apatite contains CO_3 and that the OH sites are mostly replaced by CO_3 ions. Samples, which were placed in vertical position in the SBF (vertical soaking, vs) also induced a thin CaP layer on their surfaces, as seen from the spectra. After applying the LLSI process (Fig. 23 a), the HA bands in the FTIR spectra of the samples taken out immediately after the laser irradiation (i.e. 0 hours of subsequent soaking) were hardly discernible, and only a weak band was observed at 858 cm^{-1}. This was assigned to acid phosphates. HPO_4^-containing phosphates are known to cause imperfections of the HA structure and are a precursor phase in the formation of HA. A formless peak at 1048 cm^{-1} (P-O stretching in PO_4) and a shoulder at 870 cm^{-1} (C-O bending in CO_3) were detected after 4 hours in SBF and were assigned to HA with incorporated CO_3. Further soaking for 24 hours increased the crystallinity and thickness of the layer, and revealed additional vibrations (P-O modes at 963 cm^{-1}, 1120 cm^{-1} and a well-resolved peak splitting at 1418/1454 cm^{-1}, ascribed to C-O stretching). Vibrations due to the H-O-H bending in H_2O were detected at 1645 and 3300 cm^{-1}.

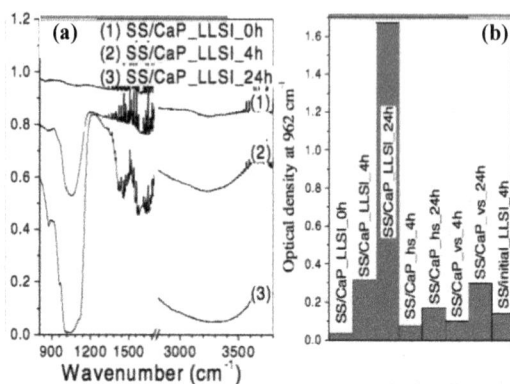

Figure 23. FTIR spectra of the CaP layers grown on stainless steel samples by: (a) the LLSI in the SBF and subsequent soaking for 0, 4 or 24 hours; (b) the histogram shows the kinetics of layer growth by the two methods applied.

A relative estimation showing the thickness of the layers grown on steel can be done by using the optical density D of the layers at the most intensive phosphate peak (i.e. the one at 960 cm^{-1}). According to the Beer-Lambert law for the absorption of the infrared light by the layer,

$$D_v = \varepsilon (v)\, c\, d, \tag{2}$$

where ε is the absorption coefficient at a given wavenumber v, c is the concentration of the absorbing molecules in the layer and d is the light path length or thickness of the layer. In FTIR reflection spectra D depends on the real and imaginary parts of the refractive index of the surface layer. In the case of strong absorption, D can correlate with the concentration of the absorbing molecules as in the transmission infrared spectra. In our case the layer grows on the substrate surface from nucleation centers to microcrystals, and more and more HA microcrystals are covering the surface with time thus forming layers (observations with an optical microscope are also performed every 24 hours). Hence, D in the FTIR reflection spectra is connected with the area increase of the HA layers which cover the substrate surface and is directly related to its thickness. The optical density D (in arbitrary units) was calculated for the layer grown on the stainless steel samples by the biomimetic or the LLSI process for various wavenumbers and the results show a similar trend (Fig. 23 b). For example, D was calculated for $v = 960$ cm^{-1} and it was found that after the biomimetic growth of the layers for 4 or 24 hours in the SBF the optical density was 0.08 and 0.17 a.u. The laser-irradiated samples immediately taken out after the LLSI process showed an optical density of 0.04 a.u. The utilization of the LLSI process together with the subsequent immersion in the SBF yielded a higher optical density of the grown CaP layers: 0.32 and 1.67 a.u. for 4 and 24 hours of soaking. This increase was assigned to a synergetic effect of the simultaneous use of the laser irradiation, the solid substrate and the SBF precursor solution. It is generally accepted that when applying simultaneously a few stimuli to a system, there is a possibility of a synergism, i.e. where the total result of all stimuli is greater than the sum of the individual effects [5].

2.   Development of an Optical Diagnostic of Laser Plasma for *In-Situ* Monitoring of Laser Irradiation During the Laser-Liquid-Solid Interraction

The LLSI resulted in an enhanced CaP formation on the material surface, compared to the biomimetic method of prolonged soaking in SBF, and this result was attained as a single one-step and time-saving process. The new laser processing method can yield a significant progress in the materials coating with bone-like apatite in terms of nucleation rate, simplicity and availability to coat complex shapes. Other advantages are that it was carried out under room temperature and atmospheric pressure, i.e. it did not require the presence of buffer gasses or vacuum conditions. However, the mechanisms of the enhanced CaP formation are not clear enough. Observation of the laser beam interaction with the substrates immersed in the SBF by an optical diagnostic technique was necessary to give more details on the mechanisms of the LLSI process. For this purpose an optical emission spectroscopy (OES) as a plasma diagnostic tool for in situ monitoring during the LLSI process was performed. The technique traditionally used in other research areas was applied to observe the emission from excited species for a first time in such complex interaction like 'laser light–solution–solid substrate'.

The optical diagnostic was carried out by OES in the SBF. A scheme showing the LLSI set-up joined with the system for optical diagnostic of laser plasma used for the in situ monitoring of processes during the LLSI is shown in Fig. 24. The laser used was a CuBr pulsed visible laser whose emission lines 510.6 and 578.2 nm were separated by filtering in order to irradiate the samples with the wavelength of 578.2 nm.

The basis of laser performance is the generation of medium energy, up to 5 mJ with pulse duration of 35 ns at a repetition rate of 19 kHz, each of which supplies energy adequate for material removal by ablation or vaporization while effectively minimizing thermal damage effects. The substrates of stainless steel, silicon and silica glass have an area of 100 mm^2 with the top surface polished. Each substrate was placed at the bottom of a reaction vessel and SBF was poured into the container with a layer thickness of about 20 mm. Doubly distilled water was used as a control solution. After each irradiation, the solution (SBF or water) in the reaction container was changed in order to reduce the influence of particles ablated from the materials on the detected signal. The direction of the observation of the emitted light was chosen to be perpendicular with respect to the laser beam axis as shown in Fig. 24. The interaction process usually lasted for not more than 5 min and no visible loss of liquid due to evaporation occurred during this time. Only 1-2 degrees increase of the temperature of the liquid was observed. In order to study the plasma produced by laser irradiation in aqueous solutions, the laser beam was directed by a $45°$ mirror toward the glass vessel containing either distilled water or SBF. The flat vessel walls were made of quartz in order to transmit the visible and UV emissions without absorption. The laser beam produced plasma at the interaction point with the solid surface (silica glass, stainless steel, silicon) as well as at the liquid-air interface. The optical emission from plasma produced by the laser beam was acquired using a system composed by a 80 mm convergent lens, and 100 cm UV-VIS optical fiber, and 1 nm resolution SM242 spectrometer (Spectral Products, US), coupled with a 2024 pixels linear CCD detector and a computer. The acquisition time of the spectrometer was 200, 800 or 1000 ms, and the minimum time resolution of the apparatus was 5 ms. The image plane of the expanding plasma was scanned with 1 mm step by the entrance slit of the optical fiber mounted on a 3D stage. Then the emission light was directed onto the entrance slit of a SM-242 CCD (CVI Laser Optics) by the optical fiber. Fig. 25 shows the peak of the laser (CuBr, 578.2 nm) during the irradiation process, at an integrating time of 5 ms.

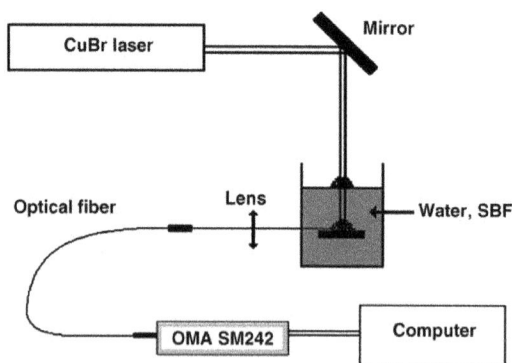

Figure 24. Scheme showing the LLSI set-up joined with a system for optical diagnostic of laser plasma used for the in situ monitoring of laser irradiation during the CaP nucleation.

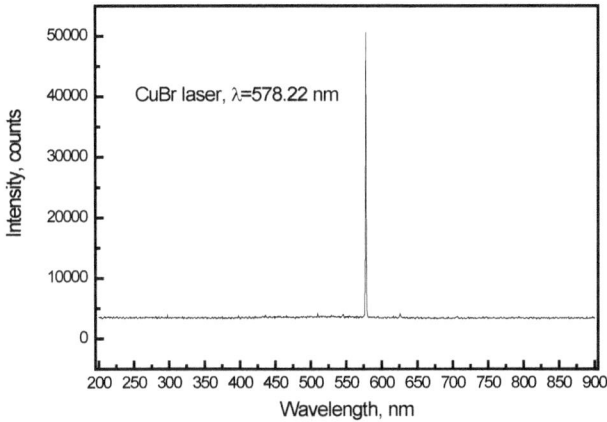

Figure 25. CuBr laser beam detected by the spectral acquisition system.

The light emission accumulated for 50–70 s (the processes analyzed in this experiment were stable within this time scale) was collected at different positions starting from the material-SBF interface and going up to the SBF-air interface to study the expanding plasma. The strongest light emission was detected at the SBF–air interface. During the laser irradiation bubbles were observed to rise above the material surface. Optical emission spectra acquired in the SBF as a liquid, and at distances of 0, 1, 2 and 3 mm above the material–SBF interface are shown in Figs. 26 and 27. Peaks were identified using standard reference data tables [32], and the National Institute of Standards and Technology (NIST) atomic spectra database [33]. Due to the fact that the intensity of the emission lines from plasma were low compared with those of the laser, larger integration time was used. Hence, due to the larger time integration, the second laser wavelength $\lambda_1 = 510.6$ nm as well as the emission lines of He I (388.86 nm) and He I (447.14 nm) were observed together with the emissions from the laser produced plasma as seen in Fig. 26.

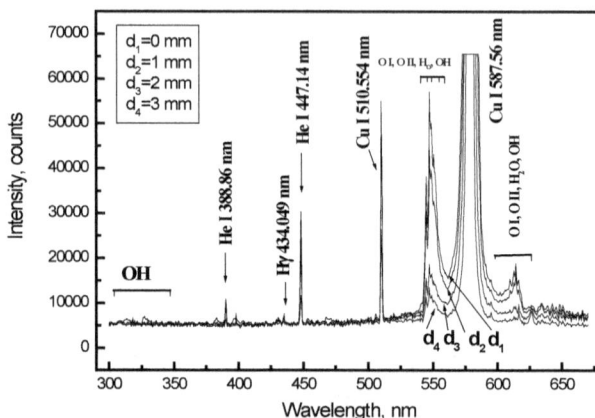

Figure 26. Optical emission spectra taken from laser produced plasma over the surface of silica glass substrate in SBF, at different distances above the interaction point.

The large peak on the right resulted from the light scattered from the laser source at 578.2 nm and the 510.6 nm line was also present in the spectra.

Analyzing the spectra in details (Fig. 27), the following molecular bands, atomic and ionization lines were identified. Peaks due to excited OH (553.06, 613.7, 623.58 and 625.8 nm) and H_2O molecule (616.57 and 618.15 nm) were detected in the detailed spectra (Fig. 27). Spectral lines of atomic oxygen O I (543.68, 595.83, 615.72 and 626.14nm) and of ionized oxygen O II (546.10 nm) were also observed. Extended spectra in several regions (not shown here) revealed additional peaks of hydrogen as follows: H at 656.27 nm, H at 486.13 nm and H at 434.049 nm, as well as He emission lines (He I at 388.86 and 447.14 nm) from the buffer gas used in the laser. Spectral lines of ionized oxygen O II at 708.39 nm and peaks in the range of 306.36–308.9 nm, at 312.17 and 347.21 nm assigned to OH molecule were detected. These are the main optical emission molecular bands and emission lines of the water elements H_2O, H_2O^+, OH, O, O^+ and H. At the chosen experimental conditions the emission lines of the SBF components were not observed. The plasma emission produced in bulk water is generally lower than at the water–air interface, with a consequent degradation of the analytical detection limit. This is due to several factors, e.g., water absorption of the laser and plasma emission and their scattering on suspended particles and micro-bubbles [34], radiation shielding by the high density plasma [35] and fast quenching in the dense medium, even for the secondary plasma from dual-pulse excitation. The complex processes involving laser-induced breakdown in water are extensively described in Ref. [36]. We consider that it will be necessary to acquire the emission signal from another direction allowing the lowest possible influence from the main laser line. In the present experiment, the acquiring direction was normal in respect to the laser beam as seen in Fig. 24. The formation of active species such as H_2O, OH, O and H can explain the high power of reaction of the plasma produced locally by the laser beam, and the possibility of CaP nuclei and particles formation just on the interaction point. Analyzing in detail the emission spectra (Fig. 28), an enhancement of the emission line intensity of oxygen was observed as the distance to the material–SBF interaction point decreased. This suggests a strong interaction of the laser beam with the atomic oxygen produced by the laser breakdown in the SBF, which led to the ionization of oxygen. We have to point out that the intensity of the oxygen lines increased only when the liquid was SBF, i.e. the presence of SBF as a liquid enhanced the production of oxygen. When distilled water was used as a control solution those line intensities were low and remained almost constant with increasing the distance from the material–SBF interaction point.

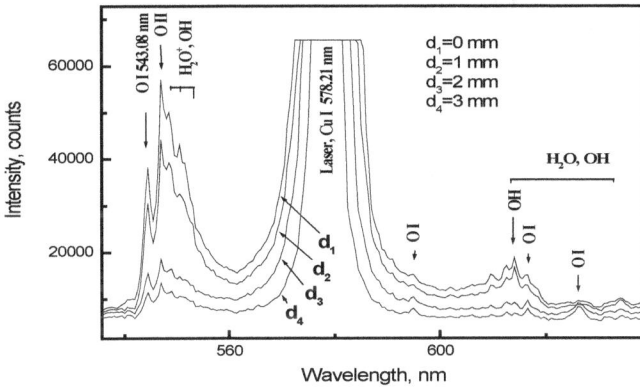

Figure 27. Expanded spectra from Fig. 27 in the range of 525 – 650 nm.

3. Cell Culture Experiments with Laser-Modified Materials

Materials, compatible with the cells are important in medical applications and therefore the interactions of the cells with the materials have been intensively investigated [37]. Cells are highly sensitive to topography, roughness, chemistry, surface charge, and hardness [38,39]. Cell–material interactions in vitro may be approximated by the process of cell adhesion and spreading, which is a convenient way to determine the biocompatibility of a material. In our study, the CaP layer grown on glass substrates by applying the LLSI process was tested in terms of its cell compatibility with osteoblast-like cells.

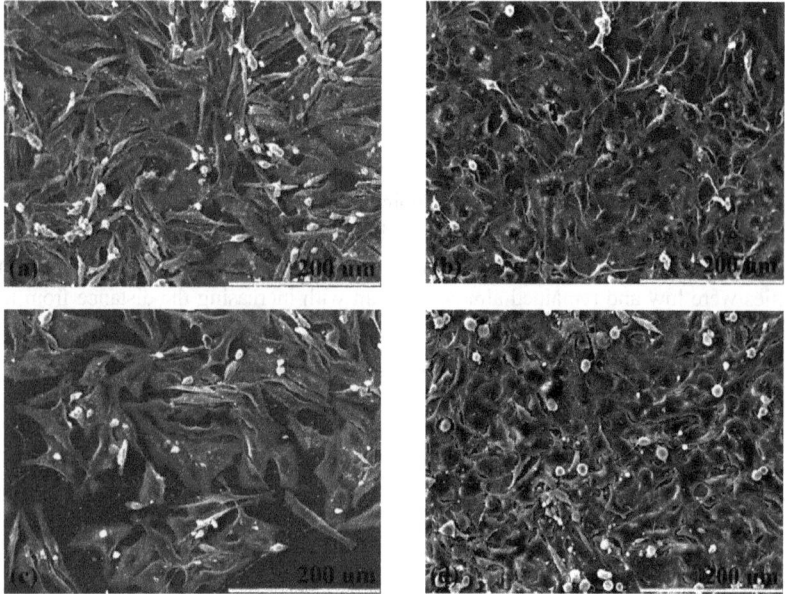

Figure 28. SEM images of osteoblast incubated on: (a,b) CaP grown on laser-treated samples for 1 and 3 days, and (c,d) on non-treated samples for 1 and 3 days.

In vitro biocompatibility of the CaP layers grown on laser-treated and control glass samples with osteoblast-like cells (MG-63, concentration of 125,000 cells/cm^2) was examined. The cells were cultured in Dulbecco's modified Eagle medium supplemented with 10% fetal calf serum and 0.5% antibiotics, incubated at 37°C and 5% CO$_2$ for 1, 3, 5 and 7 days. Osteoblasts grown in 6-well plates were prepared as controls. Cell morphology was observed by SEM (JEOL 5400 LV) after a standard procedure for cell preparation. Cell viability was evaluated by the Trypan Blue exclusion test. A Kruskal–Wallis nonparametric test was performed for statistical analysis ($p < 0.05$ was considered statistically significant). The overall cell morphology was observed using a laser scanning confocal microscope (LSCM; LSM 510, Zeiss, magnification 10x). Samples obtained by the LLSI process and controls (biomimetically coated) were covered with the protein fibronectin (FN) for 30 min to study the initial interactions of the cells cultivated for 2 hours. FN is one of the main and best characterized adhesive proteins in biological fluids, whose pre-adsorption on material surfaces may have different effects on the cell-substratum interaction and especially on its ability to induce or restore normal cell morphology and spreading. Plasma FN present in blood has been shown to promote cell adhesion, wound healing, and embryonic development despite of its low concentration in biological fluids [40].

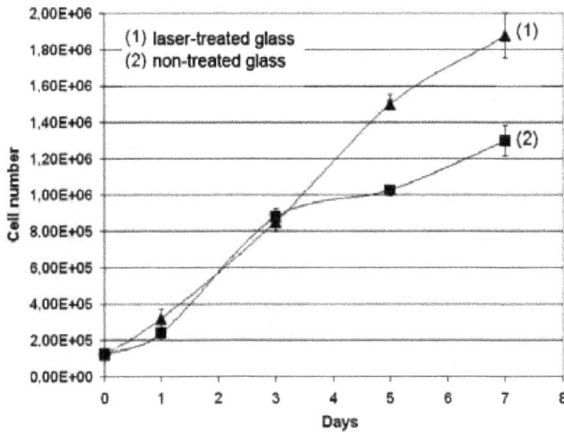

Figure 29. The proliferation results show a permanent increase in the cell number on laser-treated (1) and control samples (2) with increasing days of incubation.

In vitro biocompatibility tests revealed good cell adhesion and spreading on both laser-treated and control samples, after 1 and 3 days. As can be seen in Fig. 28 a, more cells were observed after 1 day on the laser-treated, in comparison to the non-treated samples (Fig. 28 c). The cell numbers on both types of surfaces increased after 3 days and they formed a confluent monolayer (Figs. 28 b,d). No specific cell orientation was found. Proliferation results (Fig. 29) showed a permanent increase of the cell number on the two groups of samples by increasing the incubation time. The increase was more stable on the laser-treated samples (spectrum 1). After 7 days of cell culturing, these samples showed higher cell number. A slower increase in the cell number, which resulted finally in a lower cell number, was observed for the control samples (spectrum 2).

Toxicity tests showed viability of the cells on the CaP grown on laser-treated samples of over 95%. No difference from this value was observed for the layer obtained without the laser irradiation. FN coating was found to improve the initial interactions of osteoblast cells with both types of CaP-coated surfaces (laser-treated and control) in terms of adhesion and spreading as shown by the LSCM images (Fig. 30). The number of cells increased after FN coating, and they also revealed a flattened morphology. The higher number of cells on the control FN-coated samples (Fig. 30, bottom panel) was assigned to the lower surface roughness of the CaP layer underneath.

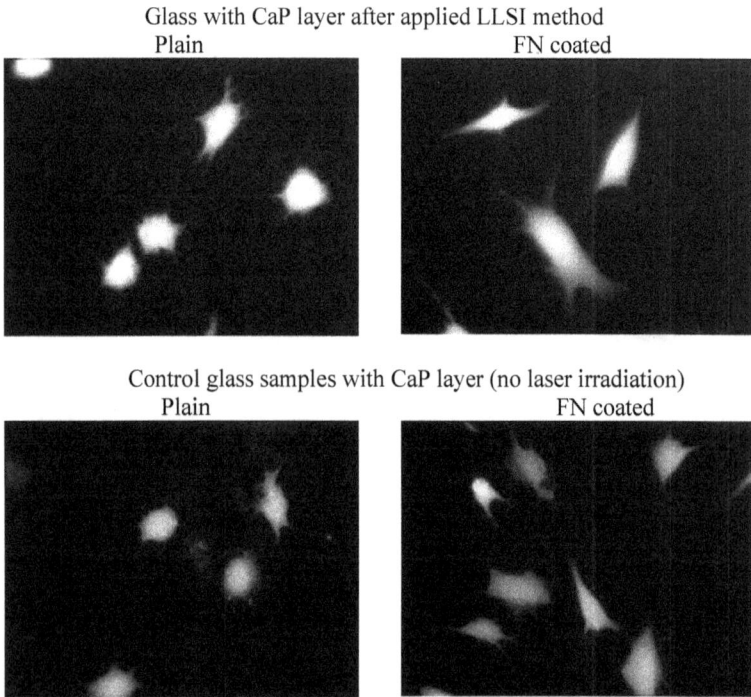

Glass with CaP layer after applied LLSI method
Plain   FN coated

Control glass samples with CaP layer (no laser irradiation)
Plain   FN coated

Figure 30. Overall cell morphology of osteoblasts adhering on laser-treated (upper panel) and control (down panel) glass samples, plain and after FN coating (magnification 20x).

4.   Summary

By applying the time-sparing LLSI method, micropatterning of various surfaces was obtained by the laser irradiation in SBF and nanostructured HA was subsequently preferably grown on the micrometer-sized areas by applying the biomimetic approach, i.e., by mineralization from a supersaturated SBF (Fig. 31). The LLSI method developed in the laboratory for Biocompatible materials of the Institute of Solid State Physics, Bulgarian Academy of Sciences is an approach that mimics natural material formation processes, and can be used to create structures organized on both micro- and nanometer-scale typical of nature. The advantage of this method is that the formation of the micrometer-scale architecture on the materials surfaces and the simultaneous nucleation of a CaP, occur within the time of the laser irradiation, i.e. less than 5 min. The nanostructured HA subsequently obtained in the SBF solution has higher thickness, ordered morphology and is grown faster than when using the biomimetic approach [1,15]. Thus, the LLSI is a controllable, precise, and time-saving process in comparison to the traditional biomimetic method for HA deposition.

The interaction of the high intensity pulsed laser beam with the materials immersed in the SBF resulted in the formation of an ablation plume of ejected material in which the surface of the material and a small amount of the surrounding solution were vaporized to form bubbles within the solution. As more material was vaporized after the multiple pulses, the bubbles expanded, until at certain critical combination of temperature and pressure they collapsed. It is believed that at this point ionization and breakdown of SBF components within the bubbles have occurred. We assume that in the LLSI process, the SBF solution trapped in the laser-induced bubbles instantly evaporated and resulted in the formation of CaP products. At the same time, the material surface was damaged and these CaP products immediately attached to the defect sites after the bubbles collapsed. Further nucleation was attained by diffusion of ions across the interfacial layer, assisted by the elevated temperature, and following supply of ions from the solution.

The applied optical diagnostic system of laser plasma shows great potential for *in situ* monitoring of laser irradiation during CaP nucleation. A strong interaction of the laser beam and the atomic oxygen produced by the laser breakdown in the SBF was observed, leading to the ionization of oxygen that was not found when distilled water was used. This technique is in its infancy, with much of the parameter space yet to be investigated. Optimizing the laser wavelength, power, fluency and focusing may improve LLSI process substantially.

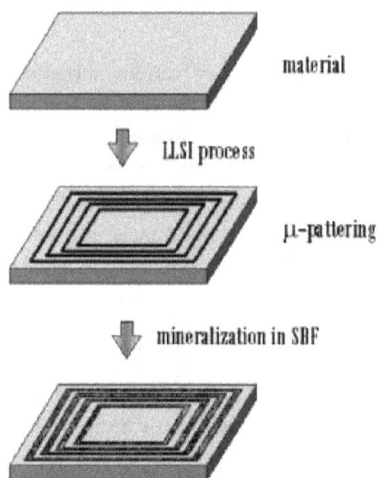

Figure 31. A schematic drawing of the LLSI method.

In vitro biocompatibility test with osteoblast-like cells revealed good cell adhesion and spreading over the CaP layer grown on both laser-treated and control glass samples, with more cells growing and proliferating on the laser-treated surfaces. Both layers grown with or without the LLSI in SBF were not toxic for the osteoblasts, which proliferated rapidly and after 3 days formed a confluent monolayer. Cell viability on the two groups of samples was over 95%. The CaP layer roughness also influences the number of adhering cells. The initial cell interactions were improved by coating the surfaces with the FN protein.

References

[1]   T. Kokubo, H. Kushitani, S. Sakka, T. Kitsugi, T. Yamamuro, Solutions able to reproduce in vivo surface-structure changes in bioactive glass-ceramic A-W, J. Biomed. Mater. Res. 24 (1990) 721-734. https://doi.org/10.1002/jbm.820240607

[2]   Y. Fang, D. Agrawal, D. Roy, R. Roy, P. Brown, Ultrasonically accelerated synthesis of hydroxyapatite, J. Mater. Res. 7 (1992) 2294-2298. https://doi.org/10.1557/JMR.1992.2294

[3]   T. Kasuga, H. Kondo, M. Nogami, Apatite formation on TiO2 in simulated body fluid, J. Cryst. Growth 235 (2002) 235-240. https://doi.org/10.1016/S0022-0248(01)01782-1

[4]   E. Lerner, S. Sarig, R. Azoury, Enhanced maturation of hydroxyapatite from aqueous solutions using microwave irradiation, J. Mater. Sci.: Mater. Med. 2 (1991) 138-141. https://doi.org/10.1007/BF00692971

[5]   J.T. Dickinson, K.H. Nwe, W.P. Hess, S.C. Langford, Synergistic effects of exposure of surfaces of ionic crystals to radiation and water, Appl. Surf. Sci. 208-209 (2003) 2-14. https://doi.org/10.1016/S0169-4332(02)01277-1

[6]   R. Sun, M. Li, Y. Lu, X. An, Effect of titanium and titania on chemical characteristics of hydroxyapatite plasma-sprayed into water, Mater. Sci. Engin. C 26 (2006) 28-33. https://doi.org/10.1016/j.msec.2005.05.003

[7]   V. Ostroverkhov, G. A. Waychunas, Y. R. Shen, New Information on Water Interfacial Structure Revealed by Phase-Sensitive Surface Spectroscopy, Phys. Rev. Lett. 94, 046102 (2005) https://doi.org/10.1103/PhysRevLett.94.046102

[8]   C. Cotell, Pulsed laser deposition and processing of biocompatible hydroxylapatite thin films, Appl. Surf. Sci. 69 (1993) 140–148. https://doi.org/10.1016/0169-4332(93)90495-W

[9]   R. Narayan, Ch. Jin, A. Doraiswamy, I. Mihailescu, M. Jelinek, A. Ovsianikov, B. Chichkov, D. Chrisey, Laser processing of advanced bioceramics, Adv. Eng. Mater. 7 (2005) 1083-1098. https://doi.org/10.1002/adem.200500155

[10]   A. Richter, P. Gonprot, R. Smith, Biofilms and their modification by laser irradiation, Nucl. Instrum. Methods Phys. Res. B 180 (2001) 1–11. https://doi.org/10.1016/S0168-583X(01)00389-5

[11]   G. Guzzardella, M. Fini, P. Torricelli, G. Giavaresi, R. Giardino, Laser stimulation on bone defect healing: an in vitro study, Lasers Med. Sci. 17 (2002) 216–220. https://doi.org/10.1007/s101030200031

[12]   A. Ebihara, B. Majaron, L.-H. Liaw, T. Krasieva, P. Wilder-Smith, Er:YAG laser modification of root canal dentine: influence of pulse duration, repetitive irradiation and water spray, Lasers Med. Sci. 17 (2002) 198–207. https://doi.org/10.1007/s101030200029

[13]   M. Kreisler, W. Kohnen, C. Marinello, H. Gotz, H. Duschner, B. Jansen, B. D'Hoedt, Bactericidal effect of the Er:YAG laser on dental implant surfaces: an in vitro study, J. Periodont. 73 (2002) 1292–1298. https://doi.org/10.1902/jop.2002.73.11.1292

[14]   A. Karacs, A. Joob Fancsaly, T. Divinyi, G. Peto, G. Kovach, Morphological and animal study of titanium dental implant surface induced by blasting and high

intensity pulsed Nd-glass laser, Mater. Sci. Eng. C 23 (2003) 431–435. https://doi.org/10.1016/S0928-4931(02)00316-8

[15] E. Pecheva, T. Petrov, C. Lungu, P. Montgomery, L. Pramatarova, Stimulated in vitro bone-like apatite formation by a novel laser processing technique, Chem. Engin. J. 137 (2008) 144-153. https://doi.org/10.1016/j.cej.2007.07.096

[16] L. Pramatarova, E. Pecheva, R. Presker, Formation of surfaces organized on both micro- and nanometer scale by a laser-liquid-solid-interaction process, Plasma Proc. Polym. 3 (2006) 248-252. https://doi.org/10.1002/ppap.200500135

[17] L. Pramatarova, E. Pecheva, Modified Inorganic Surfaces as a Model for Hydroxyapatite Growth, in Materials Science Foundations, vol. 26, Trans Tech Publications, Switzerland, 2006, pp. 1–122

[18] L. Pramatarova, E. Pecheva, R. Presker, M. Stutzmann, M. Maitz and M. Pham, Patterned surfaces for hydroxyapatite in vitro growth, J. Optoelectr. Adv. Mater. 7(1) (2005) 469-472.

[19] W. Stumm, Aquatic Surface Chemistry: Chemical Processes at the Particle–Water Interface, Wiley XIX, New York, 1987, p. 228.

[20] R. Huddlestone, S. Leonard, Plasma Diagnostic Techniques, Academic Press, New York, 1965, p. 204.

[21] L. Clark, What would happen if…? Engineering a new way to make nanoparticles, Mater. Today 2 (2000) 10–12. https://doi.org/10.1016/S1369-7021(99)80046-7

[22] T. Tsuji, T. Kakita, M. Tsuji, Preparation of nano-size particles of silver with femtosecond laser ablation in water, Appl. Surf. Sci. 206 (2003) 314–320. https://doi.org/10.1016/S0169-4332(02)01230-8

[23] D. Poondi, T. Dobbins, J. Singh, A novel laser-liquid-solid interaction technique for synthesis of silver, nickel and immiscible silver-nickel alloys from liquid precursors, J. Mater. Sci. 35 (2000) 6237–6243. https://doi.org/10.1023/A:1026701915796

[24] A. Simakin, G. Shafeev, E. Loubnin, Laser deposition of diamond-like films from liquid aromatic hydrocarbons, Appl. Surf. Sci. 154-155 (2000) 405–410. https://doi.org/10.1016/S0169-4332(99)00419-5

[25] D.A. Cremers, L.J. Radziemski, T.R. Loree, Spectrochemical Analysis of Liquids Using the Laser Spark, Appl. Spectrosc. 38 (1984) 721–729. https://doi.org/10.1366/0003702844555034

[26]   T. Bundschuh, J.-I. Yun, R. Knopp, Determination of size, concentration and elemental composition of colloids with laser-induced breakdown detection/spectroscopy (LIBD/S), Fresenius J. Anal. Chem. 371 (2001) 1063-1069. https://doi.org/10.1007/s002160101065

[27]   R. Knopp, F.J. Scherbaum, J.I. Kim, Laser induced breakdown spectroscopy (LIBS) as an analytical tool for the detection of metal ions in aqueous solutions, Fresenius J. Anal. Chem. 355 (1996) 16–20. https://doi.org/10.1007/s0021663550016

[28]   S. Koch, W. Garen, M. Muller, W. Neu, Detection of chromium in liquids by laser induced breakdown spectroscopy (LIBS), Appl. Phys. A 79 (2004) 1071–1073. https://doi.org/10.1007/s00339-004-2633-y

[29]   M. Bereznai, I. Pelsoczi, Z. Toth, K. Turzo, M. Radnai, Z. Bor, A. Fazekas, Surface modifications induced by ns and sub-ps excimer laser pulses on titanium implant material, Biomaterials 24 (2003) 4197–4203. https://doi.org/10.1016/S0142-9612(03)00318-1

[30]   E. Munoz, M. de Val, M. Ruiz-Gonzales, C. Lopez-Gascon, M. Sanjuan, M. Martinez, J. Gonzalez-Calbet, G. de la Fuente, M. Laguna, Chem. Phys. Lett. 420 (2006) 86–89. https://doi.org/10.1016/j.cplett.2005.12.048

[31]   Crystal Growth, B. Pamplin (Ed.), Pergamon Press, Amsterdam, 1980.

[32]   R.W.B Pearse, A.G. Gaydon, Identification of Molecular Spectra, Chapman and Hall, LTD, London, 1963

[33]   National Institute of Standards and Technology (NIST), www.nist.gov (last accessed November 2016)

[34]   N.F. Bunkin, A.V. Lobeyev, Influence of dissolved gas on optical breakdown and small-angle scattering of light in liquids, Phys. Lett., A 229 (1997) 327–333. https://doi.org/10.1016/S0375-9601(97)00164-3

[35]   D.X. Hammer, E.D. Jansen, M. Frenz, G.D. Noojin, R.J. Thomas, J. Noack, A. Vogel, B.A. Rockwell, A.J. Welch, Shielding properties of laser induced breakdown in water for pulse durations from 5 ns to 125 fs, Appl. Opt. 36 (22) (1997) 5630–5640. https://doi.org/10.1364/AO.36.005630

[36]   K. Kennedy, D.X. Hammer, B.A. Rockwell, Laser-induced breakdown in aqueous media, Prog. Quantum Electron. 21 (3) (1997) 155– 248. https://doi.org/10.1016/S0079-6727(97)00002-5

[37]   H. Wan, R. Williams, P. Doherty, D. Williams, A study of cell behaviour on the surfaces of multifilament materials, J. Mater. Sci. Mater. Med. 8 (1997) 45-51. https://doi.org/10.1023/A:1018542313236

[38]   L. Suggs, M. Shire, C. Garcia, J. Anderson, A. Mikos, J. Biomed. Mater. Res. 46 (1999)   22.   https://doi.org/10.1002/(SICI)1097-4636(199907)46:1<22::AID-JBM3>3.0.CO;2-R

[39]   W. Chuang, T. Young, C. Yao, W. Chiu, Properties of the poly(vinyl alcohol)/chitosan blend and its effect on the culture of fibroblast in vitro, Biomaterials 20 (1999) 1479-1487. https://doi.org/10.1016/S0142-9612(99)00054-X

[40]   D. MacDonald, B. Markovic, M. Allen, P. Somasundaran, A. Boskey, Surface analysis of human plasma fibronectin adsorbed to commercially pure titanium materials,   J.   Biomed.   Mater.   Res.   41   (1998)   120-130 https://doi.org/10.1002/(SICI)1097-4636(199807)41:1<120::AID-JBM15>3.0.CO;2-R

CHAPTER 4

Composites Based on Biocompatible Layers - Advantages and Biocompatibility Testing

Abstract

Composite materials are engineered materials created in labs and made from two or more constituent materials with different physical or chemical properties. Composites have two constituents: the major part is the matrix and the dispersed phase is only a minor part in the composite. The matrix surrounds and supports the dispersed material; the dispersed phase aims at enhancing the matrix properties. The synergism between the major and minor components produces properties, which the individual constituent materials do not possess and also allow tailoring the properties and design of composites to suit the research needs.

Keywords

Composite Materials, Matrix, Organic, Inorganic, Polymers, Nanoparticles, Proteins, Cell-Material Interaction

Contents

1.   Hydroxyapatite and Extracellular Matrix as a Second Component

The control of the mineralization processes attempted by scientists is routinely achieved in nature [1]. Organisms, such as seashells, pearls, bones, teeth and corals create a proper organic matrix to host nucleation and growth, and control solution concentrations and the supersaturation of precipitating phases. The organic molecules may either completely envelope or be enveloped by the mineral crystals [2]. They are essential for the initial deposition of mineral crystals, because they may serve as seeds for crystallization and possibly influence the number of nucleation sites generated, and also for their subsequent growth, orientation and organization. These macromolecules contribute to the strength of, and stabilize the mineral tissue. Proteins can also enhance cell adhesion and thus modulate the cellular interactions that are so important in hard tissue regeneration. An alternative theory has stated that a class of granulocytic hemocytes would be directly involved in shell crystal production in oysters [4]. Extracellular matrix (ECM) is an intracellular substance assured by the tissues, and has diverse functions, depending upon the cell type. For example: in bone, cartilage or connective tissue, ECM supports the functionality of the tissues [3]. ECM is a complex of proteins: mainly collagen, elastin, laminin, osteopontin, fibronectin etc., known as ligands (molecules that bind to another molecules) in reactions with the cell-surface receptors involved in bone physiology. Thus, on one hand, these molecules interact with the cell receptors situated in the cell membranes and, on the other hand, with the foreign surface of the material.

In bone it is accepted that bone crystals are nucleated in the interstices of an organic matrix, such as collagen fibrils, but it remains unresolved whether nucleation is activated directly by the matrix or indirectly through non-collagenous proteins absorbed in the hole zones. The minerals give the bones strength, while the ECM and particularly collagen fibers provide resiliency. An important function of the ECM proteins is to modulate the cellular interaction that is essential for hard-tissue regeneration [5]. On the other hand, the HA has been investigated as a bone substitute in order to simulate the bioactive, biocompatible, biodegradable and osteoconductive properties of natural bone [6,7]. Therefore, the composites of biomimetic HA and proteins resemble biological mineral substances more closely than do purely inorganic ones, and they are therefore considered a promising material for bone repair. The formation of organic-inorganic composites has been extensively studied [8-11]. However, subsequent deposition of HA on a pre-deposited matrix of proteins, such as a native ECM produced by osteoblast-like cells, has not been performed. From the point of view of materials used for the HA deposition, studies are usually concentrated on one type of material (i.e., metal for example) and the effects that different types of material surfaces (metals, semiconductors, insulators) have on the HA deposition is not considered. On the other hand, proteins are known to modulate the physical properties of mineral substances, and thus it is expected that they will strongly influence the structure and the biological properties of biomimetically prepared CO_3-containing HA. The aim of the work discussed hereafter was to investigate the influence of native ECM proteins, which are known as ligands in reactions with cell surface receptors involved in bone physiology, on the nucleation and growth of HA. ECM deposited on stainless steel, Si, and glass surfaces was used as a base for the HA growth.

The initial cellular interaction with biomaterials might be aproximated with the process of cell adhesion. The assessment of material biocompatibility relies heavily on the analysis of macroscopic cellular responses to the given material, which are particularly well defined with the fibroblast cell model [12]. Cell adhesion is the first cellular event that takes place on the biomaterials interface [13-17] and numerous in vitro experiments have shown that it depends on the physicochemical properties of the surface, such as wettability, surface chemistry, surface charge, roughness, etc. [18-22]. Therefore, surface modifications are frequently applied to tailor the initial tissue-biomaterial interaction [23-27].

Cell adhesion is strongly dependent on the absorbed proteins [28], particularly on adhesive ECM proteins such as collagen, FN, and laminin, which play a fundamental role due to their ability to influence the functional behavior of adhering cells. Within these proteins, FN plays a substantial role as it is the main soluble adhesive protein in biological fluids. Cell adhesion and spreading are particularly pronounced in fibroblasts, which are predominant in the connective tissue and are important in mechanisms of wounding and healing [29].

1.1 ECM/HA Composite Characterization

In order to promote interactions with bone forming cells (osteoblasts), mirror-polished AISI 316 stainless steel samples (SS; 8 x 8 x 1 mm) were prepared by coating with a native ECM (further named ECM/SS). The osteoblast-like cell line SAOS-2 (DSMZ GmbH, Germany, osteogenic sarcoma cell line) was allowed to synthesize and assemble its own ECM on the substrates under standard cell culture conditions [5]. The osteoblast-like cell line SAOS-2 is a highly differentiated, stable and non-transfected cell line, which can behave like osteoblasts. Cells were seeded at a concentration of 2×10^5 cells/ml (1×10^5 cells/cm^2) in standard alpha modified minimal essential medium (α-MEM, Biochrom) with 15% fetal bovine serum (FBS, Biochrom), kept in an humidified atmosphere with 5% CO_2 at 37°C. The cells synthesize and assemble their own ECM on substrates with sizes of 8 x 8 mm [5]. After confluency, the medium was changed with a differentiation α-MEM, now supplemented with 15% FBS, 50 µg/mL ascorbic acid, and 891 µg/mL α-glycerophosphate. After 4 days the samples were exposed to 1000 µL of 15 mM NH_4OH for 6 min for selective removal of the cells from the substrates. Subsequently, the samples were washed 5 times with 1000 µL of phosphate-buffered saline (PBS). Thus, the resulting surfaces of SS, silicon (S), and silica glass (SG) were coated with a native collagen-containing ECM with a thickness of about 0.5 µm, determined by TEM cross section of the surfaces (not given here). The matrix surface was stratified with small molecules and molecule aggregates directly attached to the surface (noncollagenous protein layer), and collagen was further embedded in the surface network, as seen by its characteristic periodicity [30]. Most of the materials' surface was covered by the noncollagenous protein layer due to the small population density of the overlying collagen fibers. Preformed binding sites having cell active structure and density were thus expected. The ECM deposited on Ti and Si substrates studied by Pham et al. revealed no significant effect of the underlying substrate on the ECM composition and deposition process [30]. Investigation with EDX and AES revealed that only N, O, and C were found after ECM deposition.

To evaluate the ability of the modified samples to induce HA growth, a kinetic study was performed by applying the biomimetic growth of HA from SBF for duration of 4 and 24 hours. To examine the mechanism of crystal growth, the samples were positioned horizontally or vertically in the solution (samples named ECM/SS_hs and ECM/SS_vs, respectively). Another group of samples was left for 4 days in the SBF. The ion concentrations were maintained by refreshment of the solution every 24 hours up to the end of the immersion procedure.

The morphology of the deposited on the ECM HA layers was investigated by SEM (DSM 962, Zeiss, Oberkochen, Germany), equipped with an EDX spectrometer. Roughness parameters (root mean square, rms) and thickness of the HA layers were determined by CPM (Leitz Linnik microscope-based system, Leitz, Wetzlar, Germany; magnification of 50x, illumination with 350-1100 nm). The structure of the layers was examined by FTIR (reflection mode, Nikolet Magna IR 750, and IRscope II, Bruker Optics IFS 66/S), Raman micro-spectrometer (HR800, Jobin Yvon Horiba, backscattering mode, 532.14 nm) and grazing incidence XRD (GIXRD; Bruker GmbH, Cu $K_{\alpha 1}$ X-ray source, λ=1.54 Å, 2θ = 20-70°, scans at 1.0° with a step size of 0.1°). The relative thickness of HA layers and the grain size were calculated from the FTIR spectra and the XRD data, respectively.

Human dermal fibroblasts were suspended in a serum-free medium and left to adhere for 2 hours at 37°C under an atmosphere of 95% air and 5% CO_2 onto four groups of samples: (1) controls of SS, S and SG; (2) ECM-coated controls (ECM/SS, ECM/S and ECM/SG); (3) samples from the second group, covered with a rough HA layer after 4 days of immersion in SBF (HA/ECM/SS, HA/ECM/S and HA/ECM/SG); and (4) control samples coated with HA after 4 or 24 hours of immersion in the SBF. Samples from the four groups were placed in 12-well tissue culture plates and studied plain or after being coated with FN (20 µm/mL for 30 min). The overall cell morphology and spreading of fibroblasts were visualized via fluorescein diacetate in living cells and observed by a fluorescent LSCM (LSM 510, Zeiss, Germany) at magnification of 10x.

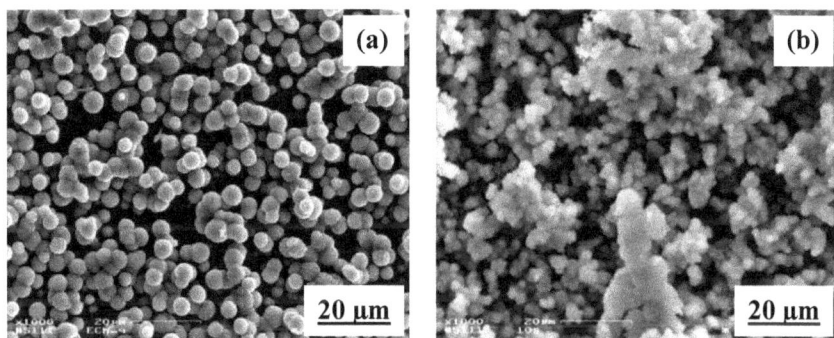

Figure 32. Morphology of the layers biomimetically grown on: (a) ECM-coated and (b) non-coated steel substrates.

Fig. 32 a shows a typical morphology of the HA layers obtained by the biomimetic method for a period of 4 days on the three materials (SS, S or SG). The SEM image shows aggregates with regular sphere shape, having average diameter of 5 µm, grouped homogeneously in a network and embedded in the ECM surface as in a matrix. This property has been ascribed to the ability of the organic molecules to control and define the crystallization process by serving as a template for mineralization, a fact observed by many researchers [2,3,8,9]. In our previous work with growing of HA layers by using the biomimetic approach, the aggregates had an irregular sphere-like shape and smaller diameters (1-2 μm), and were inhomogeneously spread on the sample surface, forming highly porous layers (Fig. 32 b; see also chapter 2). A long-distance order of the aggregates was also detected only in the case of the ECM-modified materials. Most probably the crystal nucleation was initiated and supported by the specific protein matrix laid down by the osteoblasts as suggested in [5] and especially by the noncollagenous ECM proteins, which are considered to guide the mineral nucleation. The noncollagenous protein layer generally lowers the activation energy necessary for crystal nucleation and provides nucleation sites and an oriented template for mineral deposition as observed *in vivo* [18] thus leading to modulated physical properties of the grown HA. This can be the reason why the HA grown on the ECM layers was more homogeneous and it was organized in a different way than on the control samples. This role of the ECM matrix proteins *in vivo* reported in literature was confirmed *in vitro* in our experiment.

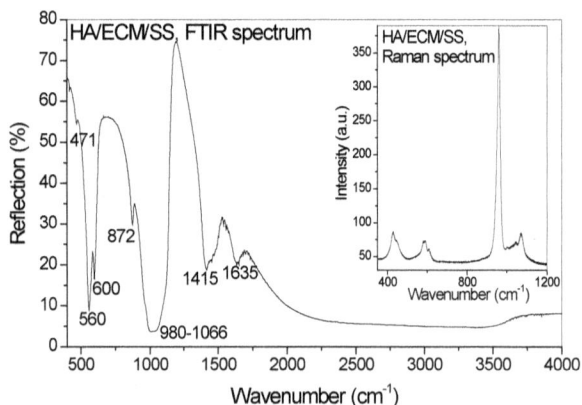

Figure 33. FTIR spectrum of HA/ECM/SS. Inset: Raman spectrum of the same sample.

FTIR spectra, recorded at different intervals of time and in the region 800 - 4000 cm^{-1} are shown in Fig. 33. They showed typical vibration modes of HA on the SS surfaces. P-O stretching vibrations (v_1, v_3) were registered at wavenumbers in the region of 962 - 1113 cm^{-1}. The incorporation of CO$_3$ was deduced from the shoulder at 835 - 876 cm^{-1}, characteristic for v_2 C-O asymmetric bending. The presence of CO$_3$ groups means that they have been incorporated in the HA structure by substituting PO$_4$ ions, which is favored at precipitation from biological solutions [3,31]. Biological HA is known as a CO$_3$-containing, low-crystalline and Ca-deficient apatite [3,31]. Carbonate groups are considered to be a principal cause of distortion in the structure of most biological apatites. The bands at 1650 cm^{-1} and 2500 - 3700 cm^{-1} in the spectra contributed to the H-O-H vibrations in adsorbed H$_2$O. Water, observed in our spectra is present in human tooth enamel and biological apatites [31]. The inset in Fig. 33 shows that in the Raman spectrum recorded from the HA/ECM/SS sample, a very strong peak at 960.4 cm^{-1} assigned to v_1 PO$_4$ symmetric stretching was present. Three groups of peaks situated under a common envelope were also observed at (428, 448 cm^{-1}), (582, 590 and 610cm^{1}), and (1045, 1072 cm^{-1}). They were ascribed to the strong peaks of v_2, v_4 and v_3 PO$_4$ vibrational modes. Two peaks at 230 and 301 cm^{-1} (an area characteristic of the crystal lattice) were assigned to Ca-PO$_4$ lattice mode [10]. FTIR spectra recorded from the HA-coated three materials were influenced by the underlying substrate. The clearest spectra presented here (Figs. 33, 34) were recorded from the SS while the S and SG did not give well resolved HA spectra as shown in details in [34]. Thus Raman spectroscopy was very useful for a complementary characterization of the HA layers since it gave structural information undisturbed by the underlying substrate, i.e. all peaks characteristic of the CO$_3$-containing HA were clearly visible [34].

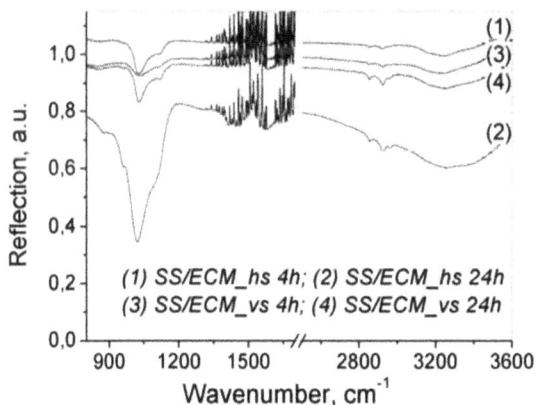

Figure 34. FTIR spectra of HA layers grown on ECM-coated SS for 4 or 24 hours in horizontal (hs) or vertical (vs) positions.

Based on the FTIR data for the HA grown for 4 or 24 hours in horizontal or vertical positions (Fig. 34), estimation of the relative layer thickness using the Buger-Lambert-Beer's law for the optical density at 962 cm^{-1} (Eq. 2) was performed according to Pecheva *et al.* [32]. The thinnest layers were observed on the samples, immersed vertically for 4 and 24 hours (vs), and samples immersed horizontally for 4 hours (hs). The calculated optical density (a.u.) was 0.05 for SS/ECM_vs_4h and SS/ECM_vs_24h, and 0.08 for SS/ECM_hs_4h. When a horizontal soaking for 24 hours was applied the optical density increased to 0.50 (SS/ECM_hs_24h). This analysis revealed that the samples soaked for 4 hours in the vertical position had a HA layer on their surfaces with a thickness comparable to that of the samples soaked for 4 hours in the horizontal position, which confirmed that the layer deposition was not simply a question of gravity. Further soaking for 24 hours did not increase the thickness of the layer grown for 4 hours, i.e., the activity of the ECM-modified surface significantly decreased and hence the velocity of HA growth decreased too.

XRD spectra showed that the HA layer grown in the presence of surface ECM proteins was thicker and with higher crystallinity [34]. The grain size calculated by using the isolated (002) reflection of the HA revealed that the crystal size in the case of the layers grown on the ECM-modified SS was higher (46 nm) than that on the non-modified samples (34 nm) confirming the lower perfection of the latter.

Ca:P ratio is characteristic of the degree of perfection and the CaP phase grown [31] and can be calculated by the EDX spectroscopy. The ratio calculated for the layers grown on the three HA/ECM-coated materials for 4 and 24 hours was similar: 1.40-1.58 and 1.50-1.60, respectively, while after 4 days of immersion in the SBF the ratio decreased to 1.30-1.40 [34]. This result means that the Ca consumption from the supersaturated SBF was higher in the first 24 hours of the sample's immersion in the solution. Later on, the P intake increased and lowered the initial Ca:P ratio, thus revealing decreased crystallinity.

The lower Ca:P ratio and the lower crystal size are equivalent to lower degree of HA crystal perfection [31]. This means that the HA layers grown after 4 or 24 hours have the highest crystallinity.

Table 4. Rms roughness of the HA layers grown on the SS, S and SG substrates after 4 or 24 hours in the SBF and on the ECM-coated substrates after 4 days in the fluid.

	HA/SS, 4 h	HA/S, 4 h	HA/SG, 4 h	HA/SS, 24 h	HA/S, 24 h	HA/SG, 24 h	HA/ECM/SS, 4 days	HA/ECM/S, 4 days	HA/ECM/SG, 4 days
rms (nm)	35	35	72	64	100	232	5×10^3	5×10^3	5×10^3

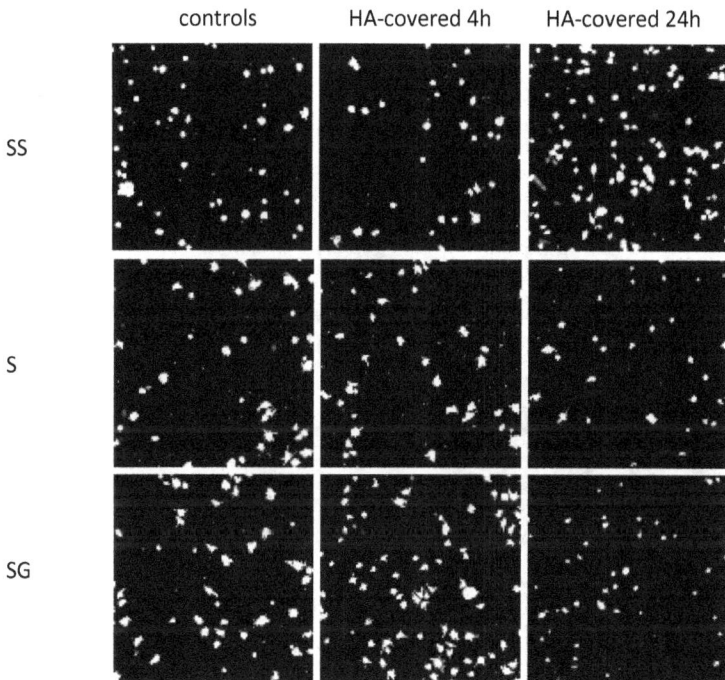

Figure 35. Overall cell morphology and spreading of fibroblasts adhering for 2 hours on plain samples (stainless steel, silicon and silica glass: controls (left panel), and after 4 or 24 hours growth of HA layer (middle and right panels); magnification 10x.

Table 4 summarizes the rms roughness of the layers grown after 4 or 24 hours and after 4 days, measured by the CPM technique. Growth of the HA layer from the SBF changed the surface roughness of the control samples, whose rms values were 7.6, 3.9, and 6.0 nm for SS, S, and SG, respectively. The rms values after growth of HA for 4 and 24 hours, increased up to approximately 200 nm. The roughest layer was formed on the SG samples after 4 hours (rms of 72 nm) and 24 hours (232 nm), followed by S substrates after 24 hours (100 nm). SS substrates after 24 hours in SBF yielded HA with an rms of 64 nm. SS and S substrates induced the growth of the smoothest layer after 4 hours of immersion (rms of 35 nm). After 4 days in the SBF, the average rms increased significantly, and it was about 5 μm for HA grown on the three ECM-coated materials.

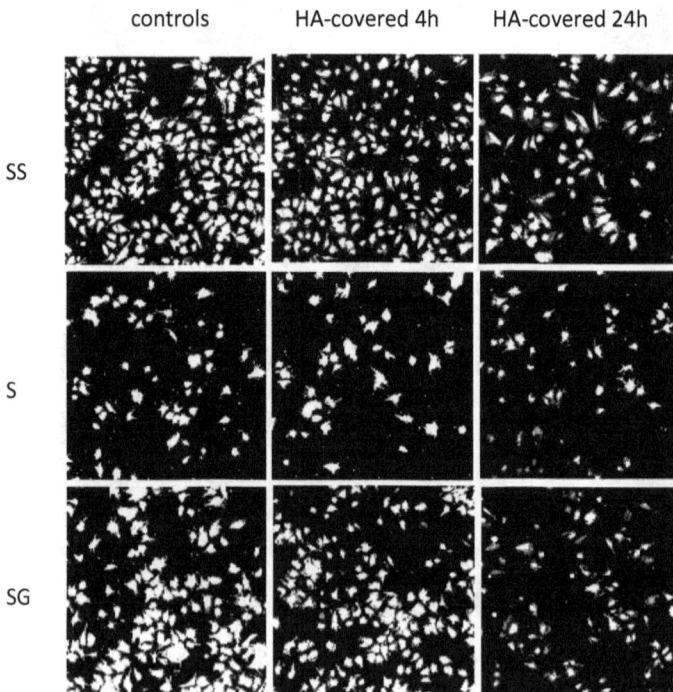

Figure 36. Fibroblast adherence on the FN-coated stainless steel, silicon and silica glass samples: controls (left panel) and with grown HA layer (middle and right panels); magnification 10x.

1.2 Initial Interactions with Fibroblasts and Effect of FN

Overall cell morphology of fibroblasts adhering for 2 hours on plain SS, S, and SG (control surfaces), as well as on the same materials but after growth of an HA layer for 4 and 24 hours, is shown in Fig. 35. Cells did not adhere or spread well on the plain control surfaces (left column), displaying a predominantly round shape. Similar cell behavior was observed after the growth of HA layers for 4 hours (middle column), although a few more cells were observed on HA-covered SG. However, after growth of the HA for 24 hours (right column) the number of adhering cells on HA-covered SG decreased and they visibly shrunk, as they did also on the HA-covered S surface. Conversely, fibroblast adhesion on the SS surface after HA growth for 24 hours significantly increased, but the cells retained their round shape. This result revealed influence of the chemistry of the underlying substrate on the fibroblast behavior.

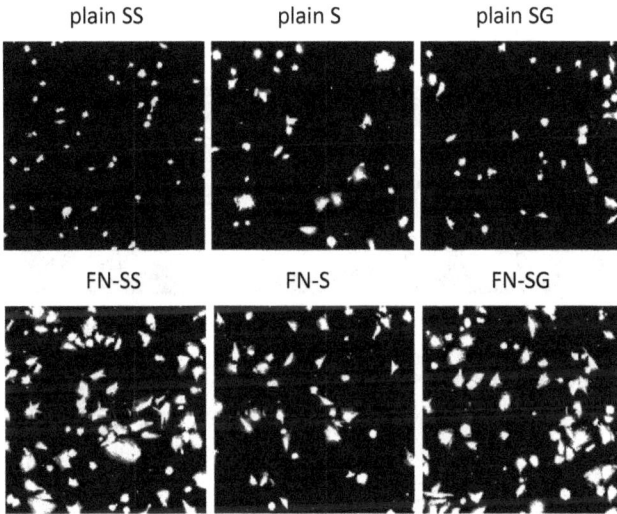

Figure 37. Overall cell morphology and spreading of fibroblasts adhering for 2 hours on controls: plain (1st row) and FN-coated (2nd row) stainless steel, silicon and silica glass samples; magnification of 10x.

The effect of FN precoating is revealed in Fig. 36. As a whole, FN improved the fibroblasts interaction with all surfaces by increasing the adhesion and restoring the characteristic fibroblast morphology, that is, an elongated cell shape and flattened morphology. This effect, however, was less pronounced on the S surfaces (control and HA-covered), implying the role of the underlying substrate. Thus, the cells on the plain S generally represented reduced adhesion, while SS yielded the highest adhesion regardless of the treatment (control or HA-covered sample). Interestingly, the highest Ca:P ratio was found on the SS substrate, thus confirming the suggestion for the chemical effect of the underlying substrate. A clear tendency for a diminished cell number with increasing time of HA growth from 4 to 24 hours was observed on the FN-coated SS and SG. This effect can be explained with the increased surface roughness of the underlying HA layers. Apart from the cells on S and SG, immersed for 24 hours in SBF, which looked slightly shrunken, fibroblasts on HA-coated SS retained their normal morphology, presumably because of the only moderate increase in HA roughness after 24 hours in SBF (64 rms) in comparison to S and SG surfaces (100 and 232 rms, respectively; see Table 4).

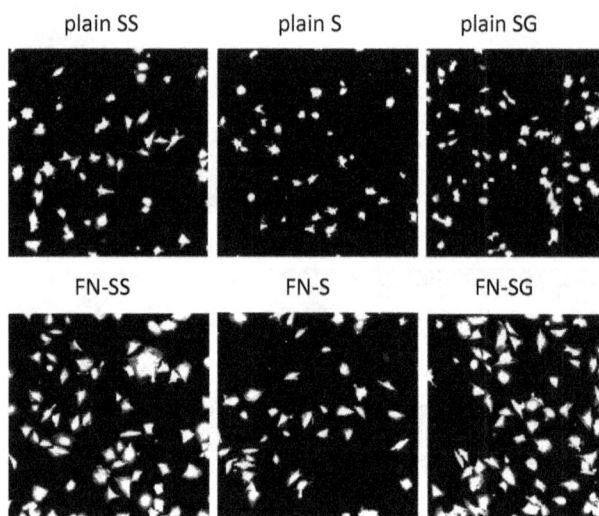

Figure 38. Overall cell morphology and spreading of fibroblasts adhering for 2 hours on ECM-coated controls: plain (1st row) and FN-coated (2nd row) stainless steel, silicon and silica glass samples; magnification of 10x.

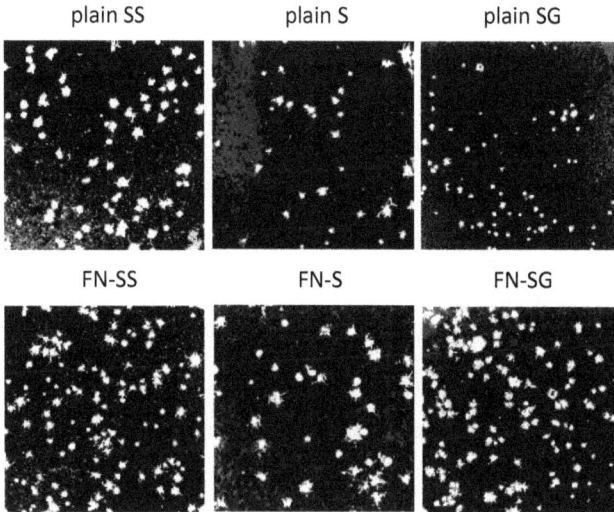

Figure 39. Overall cell morphology and spreading of fibroblasts adhering for 2 hours on ECM/HA-coated samples: plain (1st row) and FN-coated (2nd row) stainless steel, silicon and silica glass samples; magnification of 10x.

The positive effect of FN on the initial interactions of fibroblasts was confirmed in the next experiment (Figs. 37-39), where cell adhesion to the three control surfaces of SS, S, and SG was compared to ECM- and ECM/HA-coated control samples, with or without FN precoating. The samples were divided in three groups: plain control surfaces (Fig.37).

The extraction of cells also did not alter the morphology of this matrix, and thus, our results implied that it has also preserved cell attachment affinity. FN coating (Fig. 38, 2nd row) further augmented the effect of the ECM by yielding more flattened cell morphology, an effect caused presumably by the involved additional adhesive receptors that specifically recognize FN, such as $\alpha_5\beta_1$ integrins [35]. In the third group, when HA was grown on the different ECM samples by 4 days of immersion in SBF, adhesion increased only for the SS sample (first image in Fig. 39, 1st row), while the cells adhered to the same extent on the ECM/HA-coated S and SG samples as on the plain controls. As a whole, the cells did not polarize properly on all ECM/HA surfaces and developed specific stellate morphology. Here, it was difficult to distinguish the chemical or morphological effect of the underlying substrate on the fibroblast behavior since the Ca:P ratio, as well as the surface roughness of the three types of surfaces coated with HA after 4 days in SBF, were similar. The same stellate morphology was also observed when these samples were coated with FN; the number of adhering fibroblasts increased but their size remained small and the stellate morphology again predominated (Fig. 39, 2nd row). We supposed that the reason for this altered morphology in comparison to the HA layers grown after 4 or 24 hours (nanometer roughness) might be due to the extremely increased roughness of the underlying HA (rms of 5×10^3 nm; see Table 4), since the layer was grown for 4 days in this experiment, and it was found to be thicker (approximately 15 μm) and rougher (rms roughness of about 5 μm). It is known that the substrate morphology is one of the factors that influence cell behavior, and macroscopic roughness is not covered completely by cells [36], which can explain the altered fibroblast morphology. It seems that the 5 μm roughness prevented the extending of the cells shape because they sensed different heights. Presumably the small spacing that is formed between the peaks of the rough HA layer inhibited the formation of focal adhesion contacts. We consider the roughness to be the main factor that governed the fibroblasts' reactions to the different groups of samples. However, we did not exclude the possible role of the decreased Ca and P concentrations (and hence a decreased Ca:P ratio, which corresponds to decreased HA crystallinity). The positive effect of the FN coating was different for the various material surfaces in this experiment but looked more pronounced on the rougher S and SG surfaces, presumably because of the very low initial cell attachment to their plane surfaces. This fact again implies the role of the initial properties of the underlying substrate. Another reason could be that the surface energy is different from that of a smooth surface, which resulted in a lack of initial adhesion, particularly on the plain ECM/HA samples. As for the observed change in overall cell shape with increasing thickness of the HA coating, it is known that fibroblasts acquire stellate morphology when they are cultured in a 3D collagen gel system [37] which raises the possibility that a stellate morphology may appear in response to stimuli from the 3D dimension of the rough HA coating.

1.3 Ability of ECM to Influence the HA Mineralization - Summary

The three classes of materials, modified by coating their surface with native ECM proteins were able to induce the formation of HA which mimics the biological apatite in its structure and composition. However, it is more important to point out the influence of the ECM pre-coating on the layer morphology and ordering in comparison to controls (i.e. non-ECM-coated) samples: clusters with a regular spherical shape, size uniformity and higher average diameter were observed to cover the three ECM-coated materials. The clusters formed layers with a lower porosity and a very homogeneous distribution on the surfaces, and they seemed embedded in a matrix. The layer growth was not only a question of gravity settling, but also of the fact that the surfaces were activated by surface ECM proteins, so HA mineralization was induced even when the samples were placed in a vertical position in the SBF. Thus it can be summarized that the ECM proteins did facilitate the *in vitro* mineralization of biocompatible HA by serving as an *in-vitro* matrix for the mineral growth and contributing to its crystallization. This result may be successfully used in both studying the mechanism of the mineralization process in nature, as well as in tissue engineering. We consider that the non-collagenous proteins laid down by the osteoblasts in their ECM matrix had a role in guiding the mineral nucleation; they lowered the activation energy for the crystal nucleation and provided nucleation sites and oriented template for the mineral deposition as observed *in vivo*, thus leading to modulated physical properties of the grown HA. The layer grown for more than 24 hours in SBF possessed macroscopic roughness, which did not allow fibroblasts to adhere properly. Pre-adsorption of FN, as for natural ECM itself, generally improved the cellular interaction. Fibroblasts adhered with predominance on the modified SS and SG samples, which implies a chemical influence of the underlying substrate. Fibroblasts cultured on the rough HA layer acquired specific stellate morphology that resembled their shape in 3D matrices, which directed us to a predominant morphological effect. It's been concluded that the development of native ECM/HA composites on materials surfaces is an efficient way to tailor the initial cellular interaction and thus to contribute to better material-tissue compatibility.

2.   Hydroxyapatite and Nanodiamond Particles as a Minor Component

Metal materials (austenitic stainless steel, Co–Cr alloys, Ti and Ti alloys) have been widely used in medicine and dentistry due to their excellent mechanical properties and corrosion resistance in aggressive media [38]. However, their ability to bind directly to living bone is poor so they need an interfacial apatite layer to improve their osseointegration with bone. Biological apatite is Ca-deficient, poorly crystalline material, it has a Ca:P ratio below 1.67, which is the ratio of stoichiometric HA (s-HA) and low mechanical properties, and it is susceptible to ionic substitutions [39]. Therefore, the combination of the good mechanical properties of metals with the apatite bioactive properties is advantageous for application of the apatite coatings on metal implants.

Carbon-based coatings (diamond-like carbon (DLC), carbon nanotubes (CNTs), amorphous carbon, etc.) are shown to have numerous beneficial properties [40-43]. An adherent DLC coating can provide a hard, wear-resistant and corrosion resistant 'hermetic seal' for a metallic or polymeric biomaterial [40]. DLC films demonstrate outstanding hardness, very low friction and wear coefficients, and they do not offer open corrosion paths. Most importantly, tetrahedral amorphous carbon is fully biocompatible with fibroblasts, macrophages, monocytes, placental endothelial cells and neural cells, and shows absence of an inflammatory response and cellular damage [40]. DLC coatings have also an excellent haemocompatibility, expressed in a decreased thrombus formation. The extremely low friction coefficient may increase the rate of blood flow and improve other hemodynamic properties. When exposed to blood, an increased ratio of albumin to fibrinogen adsorption as well as decreased platelet activation is observed on DLC-coated surfaces. Nowadays, we have two main fields of biological applications of DLC: in blood contacting implants such as heart valves and stents (commercially available), as well as for reducing wear in load-bearing joints [42].

Several applications were proposed for CNTs, many of which are concerned with conductive or high strength composites, in which the inclusion of CNTs in a ceramic matrix is expected to produce composites possessing high stiffness and improved mechanical properties compared to the single-phase ceramic material. Remarkable increase in the hardness and slight increase in the elastic modulus of CNT-reinforced HA coating has a potential contribution toward improving bone repair [43].

Nanodiamond (ND) particles are attracting increasing interest because of their unique properties owing to their very small particle size (2–10 nm); they have high hardness, low friction coefficient, inertness to chemical attack and biological compatibility [44-49]. NDs are truly nanoscale materials; they also present the advantage of extremely high surface area, surface charge and the existence of surface functional groups (OH^-, $COOH^-$ NH_2^+ or SO_3H^-) as a result of the detonation process [9]. ND particles can be easily synthesized by detonation and they are a promising material for obtaining mechanically strong composites [47,48]. For example, ND powders can form good abrasive pastes and suspensions for high-precision polishing; ND–polymer composites are applied in hard and wear-resistant surface coatings for aircraft, cars and ships. They are considered as potential medical agents due to their high adsorption capacity, high specific surface area, and chemical inertness, and were suggested for the design of biosensors [49]. The interest in using the CNTs and nanocrystalline diamonds for medical and biological applications is also prompted by their cylindrical or spherical surface morphology which is stable with respect to cage opening under *in vivo* environments. For these applications, surface-functionalization will aid these carbon nanomaterials in becoming biocompatible, improving their solubility in physiological solutions and selective binding to bio-targets [49].

A frequently used method for coating of metal implants is electrodeposition (ED), which allows one to obtain dense apatite coatings with good adhesion to the underlying metal, uniformity, controlled thickness and deposition rate, at low temperature and under ambient conditions [50,51]. ED is suitable to produce desirable composite coatings, which gives the possibility to combine the beneficial properties of any single material used in the composite and thus to create functional materials with more comprehensive applications. Various treatment procedures undertaken after the end of the ED process lead to improvement of the crystallinity of the electrodeposited apatites.

In this study, ED technique was utilized to modify the surface of austenitic SS and Ti substrates with a biomimetic HA layer from SBF supersaturated solution. The goal was to obtain stress-free HA-ND coating with ductility and improved hardness in comparison to pure HA, by incorporating ND particles as a minor phase into the HA.

Biomimetic HA osseointegrates well due to its chemical resemblance to mammalian bone and teeth [52-54]. HA also adsorbs many proteins and interacts well with osteoblast precursor cells, therefore metal implants are frequently coated with HA in order to facilitate bone adaptation, firmer implant-bone attachment, reduced healing time and enhanced bone apposition in comparison to uncoated implants [52,53]. In order to examine the biological compatibility of the HA-ND coating deposited on stainless steel, we studied in vitro the initial interaction of osteoblast-like MG 63 cells with plain substrates, including HA-ND, pure HA and control SS surfaces, as well as after their pre-adsorption with FN and vitronectin (VN) – the main adhesive proteins in the human blood plasma. Osteoblastls are the principal cells in the bone matrix and their successful interaction with a material provides insights on its osseointegration [54]. We investigated the overall cell morphology and quantified the initial cell adhesion and spreading of MG-63 cells.

2.1 HA-ND Composite Coating Preparation

Substrates were cut from rods of austenitic SS (AISI 316L, ø15 mm) or commercially pure Ti (ASTM grade 2, ø14 mm). They were subjected to further standard metallographic polishing to obtain a mirror-finished surface (SiC papers with grit sizes #150, #320, #600, and #800, followed by alumina suspension with grain size of 0.05 μm). Finally, the samples were ultrasonically rinsed in acetone, dried under a stream of pure nitrogen (99.9 %), and stored in a desiccator.

ND particles were synthesized by the shock-wave propagation method through the detonation of trinitrotoluene and hexogen at high pressure and high temperature produced in the detonation [55,56]. Subsequent purification from graphite by applying oxidation with potassium dichromate in sulphuric acid was carried out, and after several washings with hydrochloric acid and water, the as-obtained ND powder was dried [55,56]. The purification method led to an oxidation of the ND surface, which was found to be covered with carboxyl, and to a lesser extent with carbonyl and hydroxyl groups as revealed by infrared spectra [56]. Analyses have shown that the particle size was 4-6 nm, density was 3.2 g/cm^3, growth surface was 350-400 m^2/g and the diamond content was 97-99% [56].

A potentiostat (Hokuto Denko HA150G) was used for the cathodic deposition of the coatings. The deposition was performed in a three-electrode electrolytic cell, consisting of the studied sample (SS or Ti) as the cathode, a Pt foil as the anode and a saturated calomel electrode (SCE) as the reference electrode. The SBF electrolyte resembled the ion composition, concentrations and pH of human blood plasma and was prepared as described in chapter 2 and ref. [32]. The ND particles were added to the SBF electrolyte in a concentration of 0.5 g/l and ultrasonically agitated for 20 min before the ED process (hereafter called 'ND-SBF' solution). Cathodic ED was performed with the as-prepared electrolyte, applying a potential of -1.5V (SCE) for 15, 30 or 60 min on the SS substrates or -1.8V for the same times on the Ti substrates, and maintaining the temperature at 37°C. After the deposition, the samples were washed under a flow of distilled water, dried in air and stored in a desiccator.

The morphology of the electrodeposited layers was studied by SEM (Hitachi S-3400 NX, 12 kV), coupled with EDX spectroscopy, performed over five areas with size of 20x25 μm, as well as from five points taken on white aggregates. Topography, roughness and thickness parameters were investigated by AFM (SOLVER SPM, NT-MDT, non-contact mode) and CPM (Z scan mode, Leitz Linnik microscope, 50x objective, white light illumination in spectral range of 350-1100 nm, lateral resolution of 0.45 μm, vertical resolution of 10 nm over a dynamic range of 10 μm). The layer structure was examined by FTIR (Bomem IR spectrometer FTLA2000, adsorption mode, 100 scans, resolution of 4 cm^{-1}) and micro-Raman spectroscopy (Jobin Yvon Horiba, 632 nm, resolution 2 cm^{-1}). A micro hardness tester (HMV-1, Shimadzu) was utilized to estimate the Vickers hardness (HV) by applying a load of 245 mN for 10 s and performing five indents on each sample (for the selected load and time, the effect of the substrate on the hardness can be ignored). XPS measurements (Surface Science Inc. SSX100, USA; Al Kα excitation source, E= 1486.6 eV) were performed to investigate the surface chemistry of the composites; calibration of the binding energy was done by setting the C 1s line to 285.0 eV. The mineralization ability of the composite coatings (i.e. their ability to induce new apatite formation) was tested in SBF at body temperature (37°C) for up to 7 days. For this purpose, a batch of samples was immersed in SBF (the solution was renewed every day) and each day several samples were taken out, washed with Milli-Q pure water, dried in air and analyzed by SEM, EDX and FTIR.

The XRD pattern of the purified ND powder (Fig. 40) showed an intensive diffraction line at 2θ = 44° assigned to the (111) crystalline face of diamond (C sp^3 hybridized carbon) and a low intensive peak at 75° due to the (220) face of diamond. The peaks at

25.5 and 27.0° corresponded to the presence of graphite (sp^2-hybridized carbon, (003) face) as a shell of the diamond core. The ND crystalline grain size was determined by using the Scherrer's equation for the (111) peak as being 4.5 nm. TEM image of the ND particles confirmed that the NDs are smaller than 10 nm (Fig. 41). The diffraction pattern revealed the (111), (220) and (131) crystallographic planes of the nanocrystalline diamond (inset shows the diffraction rings) and an additional plane due to graphite. Thus, it was concluded from both XRD and TEM data that the ND particles consisted of a diamond core wrapped in a graphite shell.

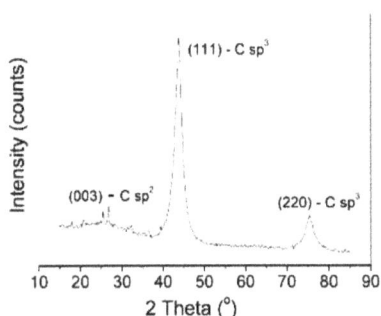

Figure 40. XRD pattern of the purified ND powder.

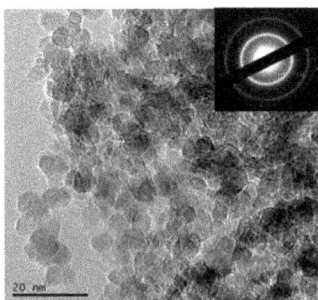

Figure 41. TEM image and diffraction pattern (inset) of the ND particles.

During the ED of CaPs from an aqueous electrolyte on a metal cathode, reduction of water and dissolved oxygen occurs at the surface of the cathode [57]. As a consequence, hydroxide ions are generated and hydrogen gas evolution arises simultaneously, according to the reactions (3) and (4):

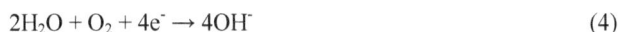

$$2H_2O + 2e^- \rightarrow H_2 + 2OH^- \qquad (3)$$

$$2H_2O + O_2 + 4e^- \rightarrow 4OH^- \qquad (4)$$

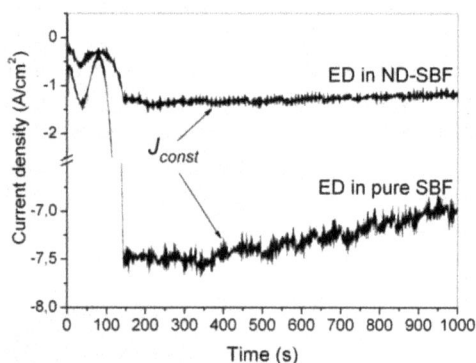

Figure 42. Cathodic current density during the ED of ED-AP and ED-AP-ND coatings.

The cathodic current in the experiments was recorded and used to calculate the current density over the specimen area (3 cm^2) exposed to the electrolyte. As shown in Fig. 42 for ED carried out for 15 min in ND-SBF and in pure SBF solutions on steel substrates, the current density increased with the initial increase of the potential in the cathodic direction and became almost constant after reaching the target potential of -1.5 V (the same behavior was observed for 30 and 60 min of ED, as well as for 15, 30 and 60 min of ED on Ti and therefore these graphs were not presented). The increase of the current density just after the cathodic charge was due to the ED of a CaP layer. During this increase water was decomposed to H$_2$ and OH$^-$ groups and gas evolution was observed. A decrease of the current density during the increase and a peak were observed for both curves in Fig. 43. The current density decrease (i.e. the minimum in the current density) was generated by the reduction of nickel in the SS. After reaching the steady-state current, H$_2$ evolution continued through the electric reduction of water. The region of steady current density (J$_{const}$) in both curves was attributed to the passivation of the metal surface due to the formation of an electrodeposited layer after 15, 30 or 60 min. The current density recorded for the samples with ND particles in the SBF was much smaller than for the samples without ND. The reason was most probably a higher electric resistance of the deposited composite layer, which worked as a protective film for the current, and consequently as a barrier for the mass transportation. After ED for 60 min on steel and Ti, J$_{const}$ for both curves showed a slight decrease, thus indicating that the deposited layer increased in thickness to such a degree that it started to isolate the metal electrode.

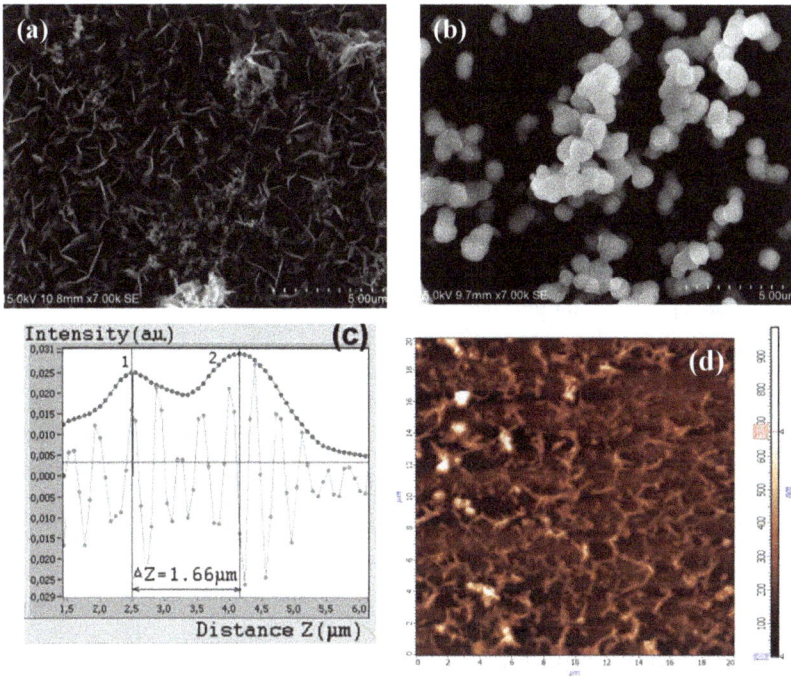

Figure 43. SEM images of (a) the coating obtained after ED from ND-SBF for 15 min on stainless steel and (b) HA coating obtained by the biomimetic approach; (c) AFM image of (a); (d) Z scan on a platelet in the ED-HA-ND coating.

The SEM image of the composite coating electrodeposited for 15 min from ND-SBF solution on SS or Ti substrates revealed the formation of a dense and uniform coating with platelet-like morphology (Fig. 43 a). White aggregates of platelets were also found, randomly distributed on the underlying homogeneous coating. Increasing the duration of the ED to 30 or 60 min did not change the coating morphology and the same platelet-like coating with attached aggregates was observed on both materials. The coating on the HA control samples obtained from pure SBF after 15, 30 or 60 min on steel or Ti had the same morphology, i.e. the incorporation of the ND particles from the SBF did not influence the morphology. For comparison, the biomimetically grown HA (i.e. without ED, Fig. 43 b) revealed the formation of non-homogeneous coating consisting of white sphere-like particles grouped in clusters as already discussed in chapter 2. AFM imaging was used for complementary information (Fig. 43 c) and it showed the homogeneously distributed platelet crystals forming a dense layer with random aggregates having lateral sizes of a few μm. Peak-to-peak and rms roughness of the coatings were about 1 μm and 99 nm, respectively. CPM technique was used as a complement to the other surface measurement techniques such as AFM and SEM to estimate the platelet height. Thus, knowing that the coating thickness was about 1 μm according to the AFM data, the height of the platelets was estimated by a Z scan along the optical axis to be about 1.66 μm as seen by the CPM measurements in Fig. 43 d where peak 1 comes from the air-platelet interface and peak 2 from the coating-substrate interface. Comparison of the EDX spectroscopy measurements performed over areas or as point measurements taken on the white aggregates on the coated SS and Ti substrates (Table 5) revealed that the platelet aggregates in the composite coating had more carbon than the coating itself (see area measurements) and then the control HA coating. Probably during the process of ED, the ND particles agglomerated in the white aggregates despite the preliminary sonication of the precursor solution. As reported in [15], ND particles produced by the detonation method easily form conglomerates with a characteristic size of the order of 1 μm because of their high surface energy. It was observed that ND particles present in water solution form conglomerates ranging from a few hundred to even a few thousand nm, depending on the solution pH. Thus, the aggregates in the composite coatings can be attributed to ND conglomerates deposited from the ND-SBF precursor solution, which corresponds well to the sizes of the observed aggregates, i.e. a few μm (Fig. 43). Apart from the high C content, Ca and P were detected in the platelet crystals forming the aggregates by EDX. Control electrodeposited coatings grown in pure SBF on steel (i.e. HA samples) had the highest Ca:P ratio (Table 5). It decreased with increasing time of deposition as follows: 1.61, 1.46 and 1.40 after 15, 30 and 60 min of ED, respectively. Further, after deposition from ND-SBF on steel, the Ca:P values of HA-ND coatings were lower than those obtained for the HA control coatings, namely: 0.83, 1.09 and 1.29 for 15, 30 and 60 min, respectively. The decrease in the Ca:P ratio in these samples could be attributed to some degree of amorphization, which the ND particles introduced in the already non-stoichiometric coatings. In addition, the Ca:P value increased with the time of ED, showing that the HA coatings matured with time [24,25]. The Ca:P ratio of the composite coatings on Ti was similar (about 1.15) for the 15, 30 and 60 min of ED from the ND-SBF solution as seen in Table 5. The coating on the HA control samples, obtained from the pure SBF after 15, 30 or 60 min also had a low Ca:P ratio. As the value for synthetic s-HA is 1.67, the lower values calculated for the control coatings could be attributed either to low crystallinity apatite, or to a combination of HA and other CaP phases, which also contribute to a decrease of the coating crystallinity [31,56]. The lower values match well with the observed morphology of coatings, consisting of platelet crystals (s-HA has a needle-like morphology [31]).

Table 5. EDX results for the Ca:P ratio and C content (at.%), measured over areas (20 x 25 μm in size) or by point measurements taken on white aggregates on the HA-ND composite and HA control coatings deposited on stainless steel and Ti substrates.

Samples	C concentration [at%] (SS)		C concentration [at%] (Ti)		Ca:P ratio (SS)	Ca:P ratio (Ti)
	Area	Aggregate	Area	Aggregate		
HA-ND15	2.50	35.45	4.57	47.74	0.83	1.15
HA-ND30	4.18	22.00	1.84	27.74	1.09	1.13
HA-ND60	5.66	8.81	11.23	49.65	1.29	1.15
HA15	3.96	12.33	1.49	1.94	1.61	1.14
HA30	5.43	9.91	0.46	0.99	1.46	1.21
HA60	4.88	11.96	1.98	2.21	1.40	1.10

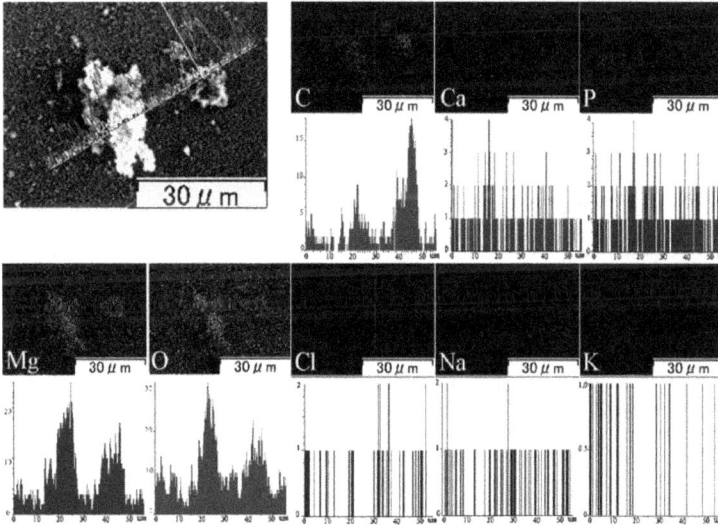

Figure 44. EDX elemental mapping over area (20x25 μm) on the HA-ND15 coating of stainless steel substrate (images) and over a line crossing the white aggregates (spectra below the images).

Apart from the Ca and P signals, K, Na and Cl were detected in the composite and control coatings in minor amounts (Na = 0.02-0.27 at%, Cl = 0.02-0.19 at% and K = 0.04-0.10 at%), independent on the substrate used for the ED and similarly to the natural bone composition [31,56]. Comparison to the HA growth by the biomimetic approach revealed a strong increase in the Mg content in the two types of electrodeposited coatings (composite and control) discussed in this section. Mg was especially accumulated in the aggregates (Fig. 44). According to the literature, increased Mg content in the case of electrodeposited apatite coatings has been usually observed. Thus, the non-stoichiometry of the HA-ND composite coating could be attributed to its high Mg content as it is known that Mg is an element which introduces imperfection in the HA lattice together with the incorporated carbonates [31,56]. EDX elemental mapping over area (20 x 25 μm) on the HA-ND15 samples (Fig. 44; SS substrates) confirmed the higher concentrations of C, Mg and O with respect to the HA samples, especially in the white aggregates, as well as uniform distribution of Ca, P, Na, Cl and K. The same results were obtained for the HA-ND coated Ti substrate.

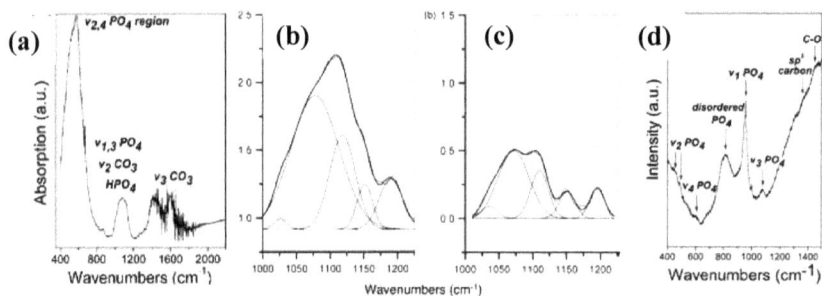

Figure 45. Representative FTIR (a) spectra of the HA-ND composite coating observed in Fig. 43, (b) peak fitting of the FTIR spectra for the ED-HA-ND60 samples and (c) for the same sample after treatment in vacuum at 750⁰C for 60 min, (d) Raman spectra of the HA-ND composite coating in Fig. 43.

The composite structure was investigated by FTIR and micro-Raman spectroscopy. Fig. 45 a shows the FTIR spectrum of the HA-ND coating obtained on steel for 15 min of ED and it is representative also for 30 and 60 min of ED of the composite. A highly intense envelope of peaks due to $v_{2,4}$ PO_4^{3-} stretching modes was detected in the frequency region of 400-700 cm^{-1}. A broad peak envelope with lower intensity was observed at wavenumbers of 950-1180 cm^{-1}. Peak fitting revealed that it was an envelope of under-lying peaks due to $v_{1,3}$ PO$_4^{3-}$ stretching, v_2 CO$_3^{2-}$ stretching and HPO$_4^{2-}$ vibrational modes. In addition, two absorption peaks due to v_3 CO$_3^{2-}$ stretching at 1430 and 1620 cm^{-1}, characteristic of partial CO$_3$ substitution of PO$_4$ ions in the HA structure were detected [31,56]. Thus, it was concluded that the coating was non-stoichiometric, CO$_3$- and HPO$_4$-containing HA. As known, incorporation of carbonate and acid phosphate is typical for the HA in bones [31]. Both ions are found to induce disorder in the s-HA thus leading to the formation of Ca-deficient and low crystalline HA, which explains the broad peaks in the FTIR spectra. The spectrum intensity of the HA-ND coatings was lower than those of the corresponding HA coating (spectra were identical and hence not shown here), thus revealing lower thickness (as known, peak intensity is proportional to the coating thickness according to the Beer's Law [58]). Probably the HA growth was retarded with the incorporation of the ND particles. The same peaks characteristic of CO$_3$- and HPO$_4$-containing HA were observed for the composite coatings on Ti substrate.

Representative coated samples were subjected to annealing in vacuum (2×10^{-2} Pa) at 750°C for 60 min. This treatment was found to improve the crystallinity of the layers as observed by FTIR spectroscopy through the increase of peak intensity and the decrease of peak width: the peaks lying under the envelope in the PO$_4$/CO$_3$/HPO$_4$ region were better resolved (Figs. 45 b,c). Decrease of the coating thickness was also detected by the decreased intensity of the peaks in the FTIR spectra.

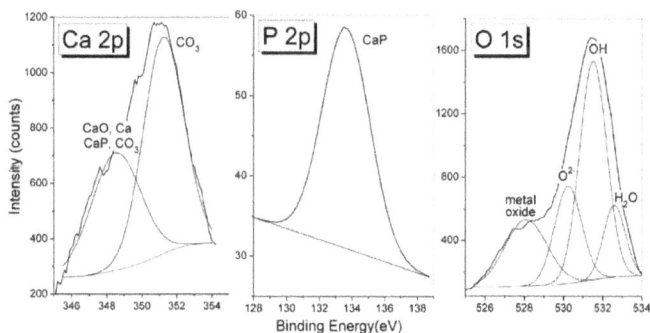

Figure 46. Deconvoluted XPS spectra of Ca 2p, P 2p and O 1s for the layers obtained by ED from the ND-SBF electrolyte.

For more detailed information Raman spectroscopy was used to study the HA-ND composite layers. The characteristic Raman spectrum of the composite deposited on SS substrate for 15 min of ED is depicted in Fig. 45 d and is representative for 30 and 60 min of ED of the composite. Vibrational modes due to P-O in PO_4, typical for HA were detected: strong v_1 symmetric stretching at 960 cm^{-1}, v_2 asymmetric bending at 440 cm^{-1} with a shoulder at 480 cm^{-1}, shoulders due to v_4 symmetric bending at 540 and 600 cm^{-1}, and v_3 asymmetric stretching at 1078 cm^{-1}. Vibrations due to P-O in a disordered continuous phosphate network at 808 cm^{-1} with a shoulder at 670 cm^{-1} [59] were found. A shoulder at 1360 cm^{-1} could be due to sp^3 hybridized carbon coming from the presence of the ND in the coatings. C-O vibrations due to CO_3 in the HA or carboxyl (COOH) groups on the ND surface were observed as a shoulder at 1460 cm^{-1} [60]. Raman and FTIR spectra of the corresponding HA-ND and HA coatings on Ti substrates were identical and thus were not shown here.

Table 6. Vickers hardness (HV, mean ± SD) of the coatings obtained by ED from ND-SBF and pure SBF solutions on stainless steel and Ti

	Vickers hardness					
	HA-ND15	HA-ND30	HA-ND60	HA15	HA30	HA60
Steel	310.0±7.8	394.0±42.6	453.8±66.7	335.0±15.3	355.0±29.2	373.7±32.3
Ti	218.2±9.9	246.0±15.9	205.0±13.4	186.6±14.6	193.6±11.1	197.4±17.1

More information on the chemical state of the composite coating was obtained by XPS. The deconvoluted XPS spectrum of Ca 2p on the surface layer obtained by ED from the ND-SBF electrolyte on Ti substrates (Fig. 46) confirmed the formation of CaP, carbonate, as well as calcium oxide (under the common envelope of Ca 2p$_{3/2}$, centered at 348.5 eV [61]). Typically, Ca 2p has a doublet structure and the second peak situated at 350.5 eV (not shown) was unambiguously attributed to $CaCO_3$ compound. CaP was also observed in the P 2p spectrum through the peak at 133.5 eV [61]. In addition, the deconvolution of the O 1s spectrum revealed peaks originating from O^{2-} at 530.2 eV, adsorbed water from H_2O at 532.6 eV and the predominant existence of OH$^-$ groups (peak at 531.5 eV), and hence a hydrated surface layer, which was also an indication of CaP formation. Metal oxides were also detected from the O 1s spectrum by a peak at 528.0 eV [61].

We also examined the coatings hardness since this parameter is a guarantee of good wear resistance, which is crucial for HA implant coatings. According to the HV results presented in Table 6, deposition of the composite coating for 15 min from ND-SBF on SS did not increase the HA hardness (HV310 versus HV335 for the HA coating). However, the composite obtained after 30 or 60 min (HV394 and HV454, respectively) showed higher HV values than the HA coating deposited for the same time from pure SBF (HV355 and HV374, respectively). Thus, despite the low initial concentration of the ND particles in the ND-SBF electrolyte and the sedimentation of the bigger particles with the ED time, the HA hardness was improved with the higher content of ND particles. HV results for the layers obtained on Ti substrates (Table 6) revealed that the deposition from ND-SBF yielded an increase of the HA-ND coating hardness in comparison to pure HA. The highest value for these samples was measured after ED for 30 min and was HV246. For comparison, the control samples deposited after 15, 30 or 60 min of ED from pure SBF, all had hardness values less than HV200. The substrate used for the ED influenced the HV values, which were lower on the Ti substrates. SAICAS measurements confirmed the AFM and CPM data on the coating thickness to be about 1 μm.

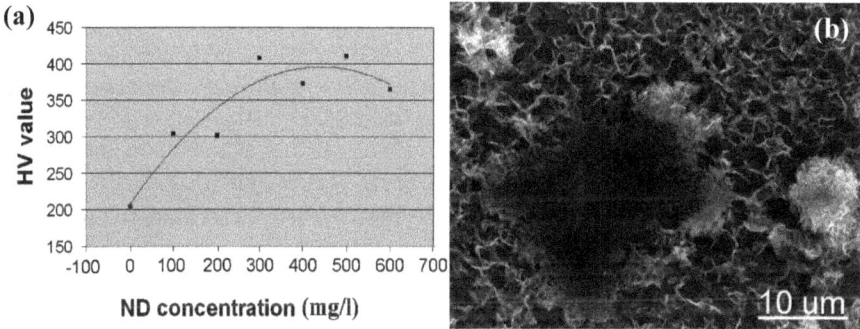

Figure 47. (a) Change of the HV values for HA-ND coatings obtained by stirring of ND-SBF electrolyte with varying ND concentrations; (b) qualitative data obtained by SEM image of the imprint left by the diamond indenter revealed no cracks in the composite coatings.

To avoid the ND particles sedimentation, stirring of the ND-SBF electrolyte was applied during the ED for 60 min and at room temperature of the electrolyte. Sufficiently low stirring speed was set in order to prevent the disturbance of the electric double layer created during the ED process, which is important for the formation of the HA coating. FTIR and SEM results showed no change in the coatings structure and morphology, and differences appeared only in the layer thickness and elemental concentrations. EDX data revealed significantly higher C concentration in the coating obtained with stirring, as well as a decrease of the Ca:P ratio. The latter showed that the higher ND concentration retards the HA maturation. Lower HV values were measured by the hardness test, although the much higher C concentration. It seemed that there must be some optimum of the C content in the HA-ND composite coatings which yields an increased hardness, and that increasing the ND concentration in the ND-SBF electrolyte does not necessarily mean that one can obtain higher coating hardness. In order to confirm these suggestions, we performed ED with stirring of the ND-SBF solution whose ND concentration was varied from 100 to 600 mg/l with a step of 100 mg (Fig. 47 a). Maximum hardness of the composite coating was measured at ND concentration of 500 mg/l. At a concentration of 600 mg/l the HV values started to decrease. This tendency was observed not only for the load of 245 mN but also at applying loads of 98, 490 or 980 mN (graphs not shown).

ED in this chapter was performed at low temperatures in order to simulate the body's environment and to avoid thermal stress at the metal-coating interface, which could lead to cracks in the coating and phase transformations. Low temperatures during ED are known to yield enhanced binding strength between substrate and coating. The growth of thin homogeneous layers as in this experiment (thickness of 1 μm) is advantageous since there is less possibility of layer delamination and crack formation, usually observed with thick HA layers (tens of micrometers). SEM images of the electrodeposited coatings obtained in ND-SBF and pure SBF solutions, independently on the substrate showed that cracks were actually absent. A representative image of the imprint left by the diamond indenter (Fig. 47 b) revealed no cracks extending from the imprint corners for the HA-ND coating, which testified to the good ductility and the lack of residual stress in the composite. No delamination of the coating was observed even at testing with the highest available load (980 mN).

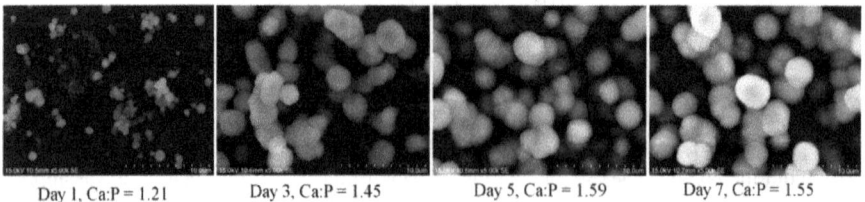

Day 1, Ca:P = 1.21   Day 3, Ca:P = 1.45   Day 5, Ca:P = 1.59   Day 7, Ca:P = 1.55

Figure 48. SEM results for the mineralization activity of HA-ND60 coatings on stainless steel (up to 7 days) revealed the overgrowth of a layer with a thickness and Ca:P ratio increasing with the time of precipitation in SBF.

To reveal the mineralization activity of the new composite coatings, i.e. their ability to form new HA layer in body fluids, they were immersed in SBF supersaturated solution at 37^0C for up to 7 days. Representative SEM images shown in Fig. 48 showed the overgrowth of a layer with a thickness and Ca:P ratio increasing with the time of precipitation in the SBF. The Ca:P ratios calculated for the layers on the 1st, 3rd, 5th and 7th day of the mineralization were 1.21, 1.45, 1.59 and 1.55, respectively. This increase revealed an improvement in the coating stoichiometry, as the Ca:P ratio steadily approached the value of s-HA of 1.67. FTIR spectra of the newly formed layers after the mineralization experiment (Fig. 49) revealed characteristic bands of CO_3- and HPO_4-containing HA with very well defined narrow peaks in the $PO_4/CO_3/HPO_4$ region, i.e. the coating crystallinity was improved. The strong peaks of CO_3 (1520 cm^{-1}) and adsorbed water (1690 cm^{-1}) observed in Fig. 45 a disappeared at the 5th day of the experiment. The same SEM and FTIR results observed for Ti suggested that the composite coating is bioactive independent on the underlying substrate and thus it is attractive as a surface modification for metal medical device materials.

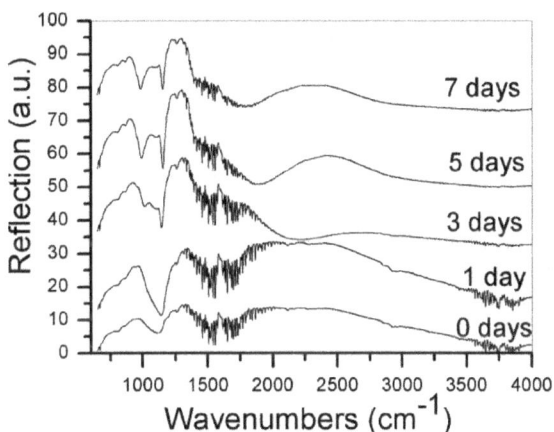

Figure 49. FTIR spectra of the HA-ND60 coatings during the mineralization experiment on steel showed that the overgrown layer was HA with improved crystallinity.

2.2   Biological Studies with Osteoblast-Like Cells

Human plasma FN (Sigma, F2006) was dissolved in 0.1 M sodium bicarbonate buffer (pH = 9.0) at 1 mg/ml, thena 10 µl of fluorescein isothiocyanate-labeled FN (FITC-FN,

Sigma, F7378) in dimethyl sulfoxide from a stock of 10 mg/ml was added. The mixture was incubated for 2 hours at 37°C. Labeled FN was separated from unreacted dye on a Sephadex G-25 column equilibrated with PBS solution. The final protein concentration was estimated by measuring the absorbance at 280 nm, while the degree of FITC-labeling was calculated against the absorbance at 494 nm. The aliquots were then stored at -20°C.

For the biological studies, the three groups of samples (HA-ND, HA and SS) were sterilized via autoclaving. For the cell adhesion studies, the three groups were coated with non-labeled FN at low (1 µg/ml) and high (20 µg/ml) concentrations, as well as with VN (1 µg/ml, Sigma) dissolved in PBS for 30 min at 37°C before washing twice with PBS. To obtain serum-coated samples, the slides were covered with pure FBS (Gibco) and further processed under identical conditions.

A human osteoblast-like MG-63 cell line (ATCC, USA) was used as model system to examine the effects of surface coatings on osteoblast adhesion, spreading and overall morphology, as well as on FN matrix organization. Cells were maintained in Dulbecco's modified Eagle medium (Gibco, 11960-044) supplemented with 10% FBS, 1% penicillin/streptomycin, 2 mM L-glutamine and 1 mM sodium pyruvate (Gibco, 11360-039), in a humidified atmosphere of 5% CO_2 in air. The culture medium was exchanged every second day. Upon reaching confluence the cells were detached with 0,05% trypsin-EDTA (Gibco, 25200-072), inactivated with FBS after detachment (approx. 5 min), then cells were re-cultured or used for the experiments.

To investigate the overall morphology, initial cell adhesion and spreading, $2x10^4$ cells/well were seeded onto each surface in 2 ml serum free medium. One set of samples was pre-coated with proteins as described above and another one served as plain, uncoated controls. After 2 hours of incubation, the living cells were stained with fluorescein diacetate (FDA) by adding 10 µl/ml from a stock of 5 mg/ml FDA in acetone. FDA can be transported across cell membranes and deacetylated by nonspecific esterases of the living cells. Resultant fluorescein accumulates within cells and allows direct visualization by fluorescent microscopy. Representative pictures of the adherent cells were taken with a fluorescent microscope (Zeiss, Axiovert 40, Germany, 20x), equipped with a digital camera. At least three representative pictures of each sample were made and morphological parameters such as the adhesion and mean spreading area of the cells were evaluated using automated image analysis software (analySIS v. 3.0, Soft Imaging System GmbH). Focal adhesions formed after 2 hours of incubation of MG-63 cells with the samples were visualized using specific mouse monoclonal antibody (Sigma) against vinculin (1:800 dilution) in 1% bovine serum albumin (BSA). It was followed by goat anti-mouse Cy3 conjugated secondary antibody (Jackson ImmunoResearch). After incubation with the antibodies, the samples were mounted and viewed on inverted fluorescent microscope and at least three representative images were obtained.

The ability of MG-63 cells to reorganize adsorbed FN (i.e. early matrix formation) was monitored on samples coated with FITC-labeled FN before seeding with $2x10^4$ cells/well in serum containing medium. After 5 hours of incubation the samples were fixed using 3% paraformaldehyde, mounted in Mowiol and viewed and photographed with a fluorescent microscope. Regular round-shaped glass coverslips (Menzel GmbH & Co KG, D 15 mm) coated with FITC-FN were used as a positive control. Late FN matrix formation (i.e. the ability of MG-63 cells to secrete and deposit FN into the ECM fibrils) was examined via immunofluorescence for FN. For that purpose $3x10^4$ cells/well were cultured on the different samples for 3 days in 10% serum containing medium. At the end of the incubation, the samples were rinsed with PBS and fixed with 3% paraformaldehyde for 5 min. The samples were then washed and saturated with 1% BSA for 15 min. Subsequently they were stained with a polyclonal rabbit anti-FN antibody (Sigma) dissolved in 1% albumin in PBS for 30 min, followed by goat anti-rabbit Alexa Fluor 555-conjugated secondary antibody (Invitrogen) for 30 minutes before washed and mounted with Mowiol.

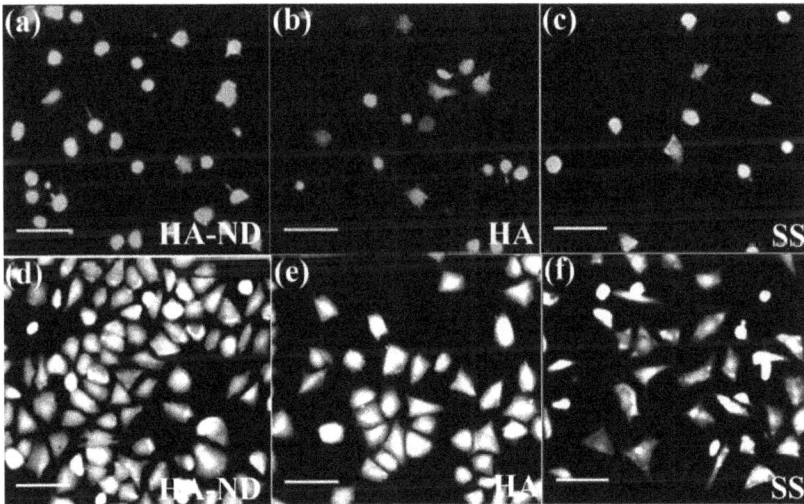

Figure 50. Overall morphology of MG-63 cells adhering for 2 hours on plain (a-c) and serum-coated HA-ND, HA or SS (d-f) samples (scale bar 100 µm).

Table 7. Cell adhesion and spreading to various samples after 2 hours of incubation in serum-free medium (significantly different values are marked with *)

Sample	Parameter studied	HA-ND	HA	SS
VN 1μg/ml	Cell number (mm^2)	163 ± 10	121 ± 19	78 ± 13
	Spreading area (μm^{-2})	1277 ± 635	1078 ± 484	1123 ± 24
FN 1μg/ml	Cell number (mm^2)	171 ± 6*	134 ± 10	86 ± 19
	Spreading area (μm^{-2})	1533 ± 697	1431 ± 603	1276 ± 582
FN 20μg/ml	Cell number (mm^2)	200 ± 14	183 ± 17	75 ± 22
	Spreading area (μm^{-2})	1963 ± 759	1836 ± 718	1682 ± 458
Serum	Cell number (mm^2)	257 ± 7*	20 ± 23	133 ± 28
	Spreading area (μm^{-2})	1414 ± 521	1284 ± 393	1142 ± 602
Plain	Cell number (mm^2)	69 ± 11	57 ± 10	70 ± 17
	Spreading area (μm^{-2})	800 ± 315	715 ± 312	673 ± 375

For the comprehensive characterization of a novel material with potential biomedical applications, it is necessary to test its biological compatibility with living cells. This was the reason to study the initial adhesion of MG-63 cells on plain and serum-coated samples after 2 hours of incubation in-vitro. The overall cell morphology on plain HA-ND, HA and SS is shown in Figs. 50 a-c. Cells attached slightly better to the HA-ND coated SS but they did not spread on non-coated samples, exhibiting predominantly round morphology. Quantitative data for cell adhesion (Table 7) showed a non-significant increase (about 20%) of the cell number on HA-ND ($p < 0.05$) (Fig. 50 a). The adhesion also did not differ with plain HA and SS samples (Figs. 50 b-c, see also Table 7). Serum coating substantially improved cell interaction to all samples (Figs. 50 d-f). Quantitative data presented in Table 7 showed that both cell adhesion and cell spreading area increased significantly in comparison with the plain surfaces ($p < 0.05$). However, adhesion of MG-63 cells was the highest on the HA-ND coating (Fig. 50 d, Table 7) in comparison with the HA and SS samples (Fig. 50 e,f, respectively). The increase of cell spreading observed on the HA-ND samples was not statistically significant ($p < 0.05$).

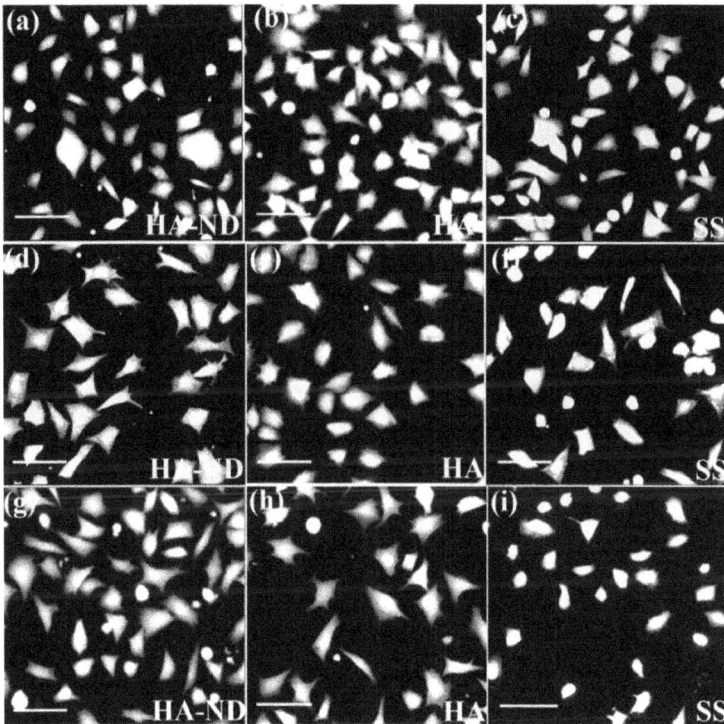

Figure 51. Overall morphology of MG-63 cells adhering for 2 hours on VN-coated (a-c), FN-coated with high (20μg/ml; d-f) or low concentration (1μg/ml; g-i) HA-ND, HA or SS samples (scale bar 100 μm).

The main adhesive proteins in serum are FN and VN. Assuming that the amount of FN is relatively low as it binds to fibrin during blood clotting, a probable reason for the improved cell adhesion to HA-ND pre-coated with serum could be the different adsorption of VN. To test this assumption, we studied the adhesion of MG-63 cells to VN in a single protein system. The samples were coated with pure VN and then the cellular interaction was studied in serum free medium. Indeed, VN promoted cellular interaction (Figs. 51 a-c), however no significant difference between samples was found, which was confirmed by the quantitative data for cell adhesion (see Table 7), a result suggesting that VN is involved in the adhesion process but this effect cannot explain the observed difference between serum coated samples. Thus, as next step we studied the effect of FN coating concentration: a saturating concentration of 20 μg/ml improved the cellular interaction equally to all studied surfaces. However, the effect appeared to be dependent on the coating concentration. Surprisingly, a significant increase in the adhesion (Table 7) of MG-63 adsorption to HA-ND coating was observed at much lower from the saturating FN concentration of 1 μg/ml, compared to pure HA and SS (Figs. 51 d-i). This result suggested either higher affinity for the FN or higher avidity of this protein on the HA-ND coatings.

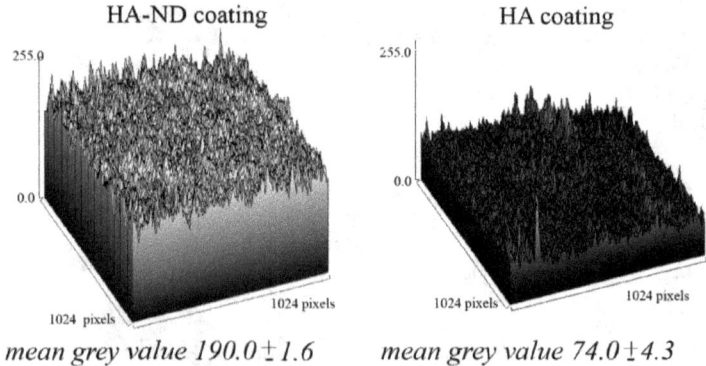

HA-ND coating   HA coating

mean grey value 190.0$\pm$1.6   *mean grey value 74.0$\pm$4.3*

Figure 52. 3D confocal images of adsorbed FITC-FN on HA-ND (a) and pure HA (b) coatings. The fluorescent intensity in grayscale pixels is analyzed to gain information for the FITC-FN adsorption.

Quantitative analysis of the fluorescent intensity in grayscale pixels based on the confocal images of adsorbed FITC-FN (Fig. 52) show a substantially higher adsorption of FN on HA-ND in comparison to pure HA, suggesting a higher affinity of these samples for FN.

The three dimensional reconstruction of the images also showed that the adsorption of FN on both HA and HA-ND coatings was rather heterogeneous, presumably following the organization of the HA crystals. As these linear features were pronouncedly expressed on the HA-ND samples, we anticipated that such topography could promote the interaction with cells [62-64]. Although no FN fibrils were observed it does not necessarily mean an absence of small micro/nanofibrils, as the characteristic scale could lie below the optical resolution. Based on these data we concluded that both FN and VN are involved in the adhesion but the improved cellular interaction with HA-ND coating should be attributed to its higher affinity for FN. Immunofluorescence study on the SS samples were not performed because of the higher refractivity of the surface.

Formation of focal adhesion complexes is another approach to learn more about the effectiveness of cellular interaction with materials. Focal adhesions are the places where real physical contact with the material surface occurs [65]. To visualize these structures, immunofluorescence for vinculin, a main cytoskeletal protein constituent of the focal adhesion complex was used. As can be seen in Figs. 53 e and f, no significant formation of focal contacts was found on plain surfaces, consistent with the low cell attachment and spreading found in Fig. 50. Nevertheless, there were more cells on the HA-ND samples, also in agreement with the improved adhesion to plain samples although they looked smaller. However, on both serum- and FN-coated samples cells acquired flattened morphology as in Figs. 50 d-f correlating with the well developed focal adhesion complexes (Figs. 53 a-d). Focal adhesions were better expressed on HA-ND samples, suggesting stronger cellular interaction on both serum- and FN-coated surfaces, particularly when compared to non-coated samples, thus revealing the important role of the adsorbed FN. Based on the data acquired, it was concluded that both FN and VN were involved in the adhesion but the improved interaction with cells of HA-ND coating can be attributed to its higher affinity for FN.

It is well documented that many type of cells such as fibroblasts and endothelial cells not only attach to adsorbed FN but also tend to rearrange it in a fibril-like pattern, presumably as an attempt to organize their own matrix [66-68]. As some surfaces can promote FN reorganization which relates positively to their biocompatibility [67], while others cannot, we decided to learn more about the adsorbed FN on the studied samples. For that purpose we cultured the cells for 5 hours on FITC-FN coated SS, HA and HA-ND, as well as on control glass samples as a positive control. As seen in Figs. 54 a-c, cells on HA-ND rearranged FN particularly well in an almost indistinguishable pattern from the glass control. FN removed by osteoblasts appeared as dark streaks on the otherwise bright fluorescent background of adsorbed FN, and this FN was further organized in a linear matrix-like structure located above or beneath the cells. When fluorescent and phase contrast images were merged, more than 90% of the reorganized FN was associated with osteoblasts, which was in accordance with previous studies [66-68] and confirm the cellular dependence of the process. This result also suggested that FN is relatively loosely adsorbed on the HA-ND coating, at least in such a way that the cells can easily remove and reorganize it in a matrix-like structure, which we consider was beneficial for its biocompatibility [66-68]. In contrast, both HA coating and SS surfaces (Figs. 54 b,d) represented only dark streaks of removed FN and only small patches of rearrangement, indicating reduced matrix-forming activity. It has to be pointed out that this effect might be relative, e.g. caused by lowered background of adsorbed FN.

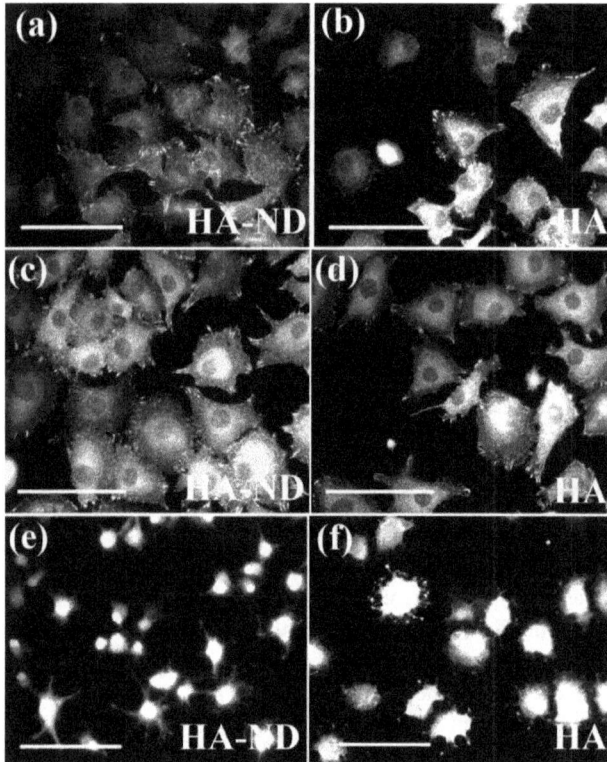

Figure 53. Focal adhesion formation of MG-63 cells adhering for 2 hours on HA-ND (a,c,e) and pure HA (b,d,f). Samples are coated with serum (a,b), with FN (c,d) or studied as plain (e,f) (scale bar 100 μm).

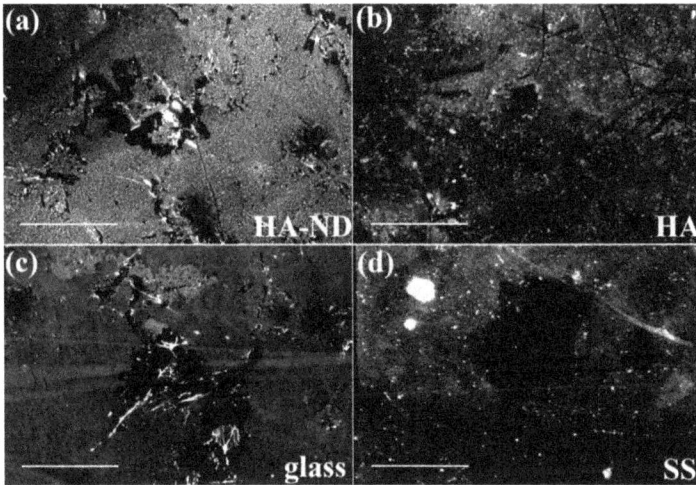

Figure 54. Early FN matrix formation on HA-ND (a) and HA (b) coatings compared with control glass (c) and SS (d) surfaces (scale bar 100 μm).

At longer incubation osteoblasts produce their own osseogenic matrix [69] containing collagens, proteoglycans and FN [70]. To learn whether the initial difference in FN adsorption will affect the subsequent development of FN matrix we cultured MG-63 cells for 72 hours and visualized FN via immunofluorescence. Multiple bright fibrils forming networks were observed on all samples (Fig. 55), indicating good cell functionality and a rather materials-independent mechanism for matrix organization. However, a trend for stronger deposition of FN matrix on HA-ND samples (Fig. 55 a) might be assumed, when comparing to plain HA (Fig. 55 b) or SS surfaces (Fig. 55 c).

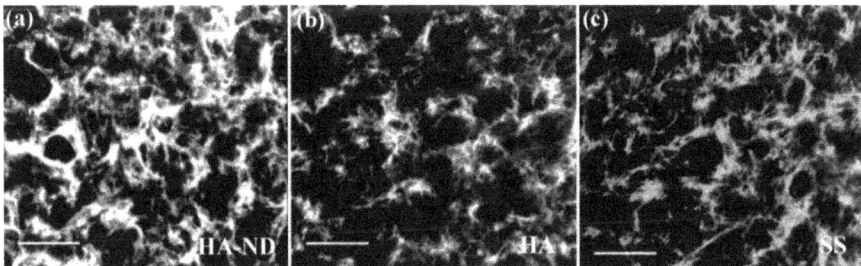

Figure 55. Late FN matrix deposition on HA-ND (a), HA (b) and SS (c) samples (scale bar 100 μm).

2.3 Ability of ND Particles to Influence the HA Properties - Summary

A dense and homogeneous, non-stoichiometric HA-ND composite coating, containing CO_3 and HPO_4 groups was obtained by ED on austenitic SS and Ti. The non-stoichiometry of the composite was assigned to the presence of CO_3 and HPO_4 ions in the coating structure, as well as to the high Mg content resulting from the ED process, which are known to introduce disorder in the HA structure. The ND particles additionally retarded the transformation of the coating to more crystalline apatite, which can be positively influenced by applying thermal treatment in vacuum. The HA-ND composite was characterized by ductility and better Vickers hardness, as well as lack of residual stress and cracks in comparison to pure HA (i.e. without NDs). A mineralization experiment in the SBF revealed that the composite coating was potentially bioactive since it induced the formation of biomimetic HA with improved crystallinity. The overall cell morphology of osteoblast-like MG63 cells and focal adhesions formation revealed improved cellular interaction on the HA-ND composite coating deposited on stainless steel resulting presumably from the increased FN adsorption. This fact combined with the sufficient organization of FN matrix by osteoblasts was another indicator of the good biocompatibility of the HA-ND coating in an osseogenic environment. Thus, the ND-reinforced HA coating can be considered as an attractive surface modification for metal materials with biomedical applications.

3. Plasma Polymerized Hexamethyldisiloxane and ND Particles

The identification of biomaterials that support appropriate cellular attachment, proliferation, and functions is critical for tissue engineering and cell therapy. There is a growing interest in functional organic/inorganic composites where a small amount of nanometer-sized material yields better physicochemical properties for cells to attach, grow, and differentiate. Composite materials are of interest because they combine the advantages of different materials to achieve specific structural properties, while a single type of material usually cannot provide all necessary properties [71]. From the biological point of view, it makes sense to combine polymer and inorganic compounds to fabricate biomaterials for bone tissue engineering, because human bone tissue is a biologically and chemically bonded composite of inorganic HA embedded in an organic matrix of collagen and noncollagenous proteins [72,73]. The first step toward the development of composites for bone substitution is the identification of the relevant class of biomaterials. A variety of natural and synthetic polymers are now available for bone tissue engineering applications but all of them have some deficiencies [74,75]. Synthetic polymers have gained a significant advantage over naturally occurring polymers because they can be produced under controlled conditions and therefore their properties are in general predictable and reproducible [76,77]. A class of synthetic polymers widely used in biomedical applications is organosilicone polymers due to their excellent inertness, flexibility, smoothness, and thermal and oxidative stability [78]. Organosilicones have been used for the production of oxygen masks, teats for baby bottles, tubes for extracorporeal circulation in heart surgery and dialysis, drains and catheters, prosthesis, contact lens, insulation coating for leads and circuits, and protective sheaths for pacemakers [79]. The low mechanical stability of organosilicones limits their application as heavy load-bearing bone substitutes. However, they can be used for deposition of thin coatings onto bone implants to improve cell-contacting properties of implants' surface. For the preparation of such coatings of great interest are organosilicones, obtained by plasma polymerization. Plasma polymerization allows deposition of high-dense, pinhole-free, and well-adherent films on a variety of substrates like conventional polymers, glass, and metals. Other advantages of the plasma polymerization process include the easy variation of process parameters and the use of modificators and fillers to produce new materials and composites with desirable properties [80]. Plasma polymerization is a process of obtaining ultra-thin films with different chemical structure and properties compared to conventional polymers and these polymers neither dissolves nor swells in a cell culture medium [81]. This is due to the monomer fragmentation and recombination in plasma. These properties depend not only on the chemical structure of the monomer but also on the plasma parameters. By varying different parameters, polymers with different chemical composition and structure could be obtained from one monomer. On the other hand the fragmentation of the molecules in the plasma leads to a similarity of structure and properties of these films despite the specific monomer structure. The plasma polymers usually have a highly branched and crosslinked structure, resulting in excellent adhesion to almost every substrate. They are pinhole-free homogeneous, chemically resistant, thermally and mechanically stable. These properties allow for a large amount of biological components to be loaded onto the surface of the films because of their biocompatible characteristics. This is why plasma polymers have been the focus of extensive research in recent years. They have potential for wide applications for tissue engineering, regenerative medicine, implants, stents, biosensors and other medical devices [82-87].

On the other side, a member of the nanocarbon family, detonation nanodiamond (DND), has emerged recently as a novel promising material for biological applications [88-94]. The nanoscale diamond material is chemically robust, nontoxic at both cellular and organism levels, and easily functionalized with different macromolecules [95-97]. Therefore, nanodiamonds can be used as reinforcements or additives in various mate-rials to increase mechanical stability and to improve tissue interactions [98–100]. One advantage of nanoparticles as polymer additives compared to traditional additives is that the loading requirements are quite low, meaning that a small amount of nanoparticles is necessary to alter properties of materials. The inclusion of only a few percent of nanosized diamond particles into a polymer matrix may increase the stiffness and strength of the polymers and may also create nanotopographic features that mimic the nanostructure of bones. The properties of polymer–DND composites can be easily tailored by changing the type, concentration, and size of nanoparticles. However, the incorporation of particles into the polymer matrix strongly influences the bonding between particles and polymers and thus the properties of the obtained composite layer [101]. Currently, there is not enough information concerning how the incorporation of DND particles into a siloxane matrix affects the surface properties of siloxane–DND composites and how this can be used to control the cell behavior for the purpose of bone tissue engineering.

In this section, the preparation of polymer/ND composite layers based on hexamethyldisiloxane (HMDS) and DND particles, embedded into the polymer matrix is described and their behavior with living cells using the osteoblast cell line MG-63 is characterized. Some studies have demonstrated that the hydrophobic surface of the polymer could be altered into a hydrophilic surface by means of O_2 plasma treatment [102]. Modification of organic polymers by O_2 plasma treatment induces some hydrophilic groups into the surface but do not change surface topography and the composition to a great degree. Other studies comparing four different plasma treatments of the polymer layers (Ar, O_2, N_2 and NH_3) found that NH_3 plasma is more effective in increasing adherence of HMDS to various materials [103]. Therefore, an aim of this investigation was also to study the effect of ammonia plasma treatment on the biological performance of the polymer films with osteoblast cells.

3.1 Synthesis of Polymer-ND Composite Layers

The polymer layers were synthesized on commercially available cover glass slides (CG, Menzel Glaeser, diameter of 15 mm), cleaned in advance by repeated washing in doubly distillated (DI) water, followed by ultrasonic treatment in acetone (99.99% purity) for 15 min, rinsing with boiling isopropanol (99.99% purity) for 10 min and then rinsing again with DI water (about 10 times). Finally, the samples were dried under a stream of pure nitrogen (99.9 %) and loaded into the reactor chamber where they were also plasma cleaned before the polymerization. HMDS $\{(CH_3)_3\text{-Si-O-Si-}(CH_3)_3$, Merck, purity > 99%\} was used as a monomer without further purification. The polymer films were obtained by plasma polymerization of HMDS monomer in vacuum chamber for 10 min under following technological conditions: 0.5 mA/cm^2 current density and 1 l/h monomer flow rate. By using several polymerization regimes it was found that the deposition rates were proportional to the polymerization time. The thickness of the plasma polymerized HMDS (PPHMDS) layers was approximately 160 nm determined by a Dektak Stylus Profile System (Vecco Instruments Inc., Woodbury, NY, USA). In order to obtain polymer surfaces with different wettability, PPHMDS films were further modified by ammonia (NH_3) plasma treatment, varying the treatment time from 30 sec to 10 min in the vacuum chamber, at the same conditions as the monomer polymerization process.

The DND powder was obtained from detonation soot, produced by the Space Research Institute of the Bulgarian Academy of Sciences (Sofia, Bulgaria), with a subsequent purification from non-diamond carbon and metal impurities through oxidation with potassium dichromate in sulfuric acid and refinement with HNO_3 and HCl [55,56,104]. Three types of DND particles were used: (i) pure DND, (ii) Si-DND and (iii) Ag-DND. The silanization of the DND surface (Si-DND) through an attachment of trimethylsilyl groups was done by mixing of a dehydrated DND powder with ethyl acetate containing hexamethyldisilazane and trimethylchlorosilane and the aim was through the silanization to prevent formation of DND aggregates [104]. After the reaction, the excess of the reagents was removed by microwave heating of the powder in butyl acetate medium. The side-produced ammonium chloride and the excess of butyl acetate were removed through threefold treatment/decanting with methanol. Si-DND sample was finally microwave dried [104]. Ag-DND powder was prepared using the ammonia complex of silver ($[Ag(NH_3)_2]^+$) added to a DND suspension under constant stirring at room temperature followed by adding of a dextrose solution. The temperature of the mixture was raised to 50 °C until the Ag incorporation into the DND surface was finalized. The powder was subsequently dried. The incorporation of Ag ions into the polymer may lead to polymer films that are highly efficient against bacterial colonization and allows the adhesion and proliferation of mammalian cells [105,106].

For the preparation of DND/PPHMDS composite layers, plasma deposition from a mixture of the HMDS monomer and the three types of DND particles in a concentration of 1 g/l was carried out. The preliminary prepared mixture was shaken for 15 min in an ultrasonic apparatus and was further stirred continuously at 275 rpm and room temperature [100]. In order to influence cellular interactions, a group of samples with the composite polymer layer were subsequently hydrophilized by plasma treatment in NH_3 atmosphere in the polymerization chamber for 10 minutes. For cell experiments the materials were sterilized in 70% ethanol for 10 min and then rinsed with sterile PBS solution.

FTIR spectra of the PPHMDS and DND/PPHMDS composites were registered by a Brucker Vector 22 FTIR spectrometer at ambient temperature in the range of 400 to 4000 cm^{-1}, using OPUS software, an average of 128 scans and a resolution of 2 cm^{-1}. UV spectra were recorded on UV-Vis spectrophotometer (JASKOV-650 double beam model with single monochromator) in 200–800 nm wavelength range at room temperature. SEM images of the layers were recorded in Carl Zeiss using backscattered electrons. The ellipsometric measurements were performed by a Woollam M-2000DI rotating compensator ellipsometer. The topography of the composites was imaged by AFM (Solver Pro; NT-MDT, Russia) and wettability parameters were calculated by using the sessile drop method under ambient condition (Easy Drop FM40 Kruss, Germany, DSA1 software; five different positions on the sample surface were measured).

Representative AFM images of the PPHDSM and DND/PPHDMS layers are shown in Fig. 56. Quantitative roughness analyses revealed that both layers had a comparatively smooth surface with R_a values of 10.4 and 15.0 nm, respectively (Table 8). Ammonia treatment smoothed the surface of the DND/PPHMDS layers even more, while the Ag-modified DND particles increased the surface roughness of the composite polymer (Ra was 50.4 nm). Water contact angle (CA) measurements revealed that the PPHMDS and DND/PPHDMS polymers had hydrophobic surfaces (CA = 81° and 84°, respectively) while the ammonia treatment or the Ag-DNDs yielded a decrease of the CA to 67^0 and 75°, respectively. Varying the time of the ammonia treatment from 30 sec to 10 min yielded a rapid decrease of the CA from 81° to about 60° at the 5th minute of the plasma treatment and increasing the duration more did not produce any further changes in the CA of the PPHMDS layer. This result showed that the polymer became more hydrophilic due to the increased amount of amino groups generated on its surface till the 5th minute. After 10 min of plasma treatment, the CA reached a plateau and did not change with time anymore revealing that the surface was saturated with ammonia groups.

Table 8. Roughness and wettability of plasma polymerized layers

Samples	R_a roughness, nm	Contact angle, ^{0}
PPHMDS	10.4	81
NH$_3$ - PPHMDS	9.8	67
DND/PPHMDS	15.0	84
Ag-DND/PPHMDS	50.4	75

PPHMDS   1-DND/PPHMDS

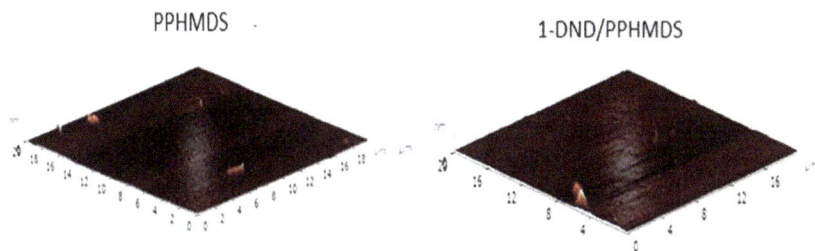

Figure 56. AFM images of the PPHMDS and DND/PPHMDS layers.

FTIR spectra (Fig. 57) showed the formation of PPHMDS by the following characteristic vibrations (spectrum 1). An increase in the intensity of the Si–O asymmetric stretching band and the appearance of a doublet (1060 and 1040 cm^{-1}) were observed and were related to the yield of a highly cross linked Si-O-Si network and a contribution from the CH_2 groups from Si-CH_2-Si bonds, formed after hydrogen subtraction. A shift and a decrease in the intensity of the bands due to the CH_3 rocking in Si$(CH_3)_x$ at 600 cm^{-1} and the disappearance of the CH_3 groups in the region of 2900 cm^{-1}, undoubtedly showed that some methyl groups were removed from the monomer molecule during the plasma polymerization. Furthermore, the changes in the polymer spectra due to methyl vibrations were an indication of the decrease of Si-$(CH_3)_3$ end groups and methyl groups in the polymer structure as a whole. In addition, the deformation mode of CH_x (x = 1,2,3) groups around 1500 cm^{-1} and the carbonyl functionalities incorporated into the polymer layers (peaks in the region 1500-1750 cm^{-1}) strongly increased, as a result of monomer rearrangement. The band shifts were related to a change in the local environment of Si-O-Si groups, which vary with the polymerization conditions, as the Si-O-C bonds exist together with Si-O-Si ones [107,108]. A change of the PPHMDS polymer characteristic bands was observed in the spectrum of DND/PPHMDS layer (spectra 2-4), which could be explained with the penetration of the DND particles in the polymer matrix.

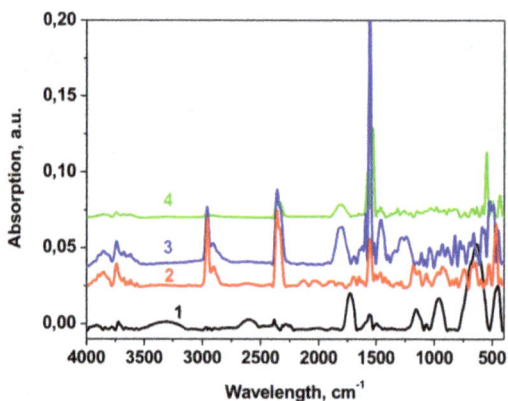

Figure 57. FTIR spectra of (1) PPHMDS; (2) DND/PPHMDS; (3) Si-DND/PPHMDS; (4) Ag-DND/PPHMDS.

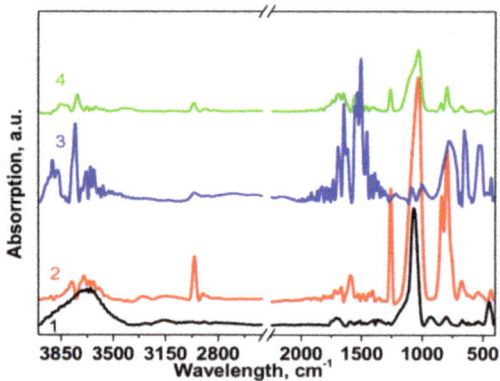

Figure 58. FTIR spectra of (1) PPHMDS/NH₃; (2) DND/PPHMDS/NH₃; (3) Si-DND/PPHMDS/NH₃; (4) Ag-DND/PPHMDS/NH₃.

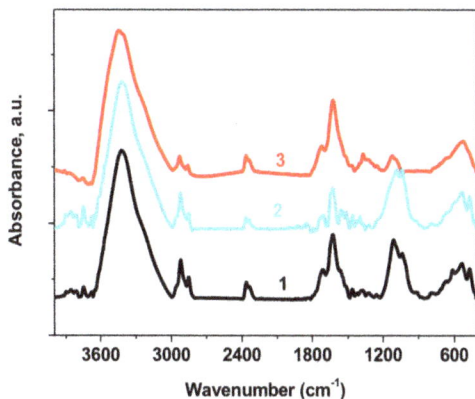

Figure 59. The FTIR spectra of the DNDs powders used as a filler of PPHMDS: (1) DND; (2) Si-DND; (3) Ag-DND.

Characteristic peaks of sp^3 and sp^2 carbon in the DND were clearly observed at 1550 and 1160 cm^{-1}. The surface of DNDs/PPHMDS was characterized with functional groups of the type –OH, >C=O, Si–O–Si, Si–O–C (Fig 58, spectrum 2). The formation of new (C-Si-C) bonds and different tetrahedral CH_3 (the intensive bands in the region 2800 – 3000 cm^{-1}) revealed the presence of both hydrophobic and hydrophilic centers on the DNDs/PPHMDS surface. In the case of Si-DND/PPHMDS (Fig 58, spectrum 3), the peaks at 500, 1200- 1300, 1470 and 1600 cm^{-1} pointed out for an increased content of amorphous carbon due to the silanization of the DND particles. The appearance of an intensive peak at 550 cm^{-1} in the spectrum of Ag-DND/PPHMDS (Fig. 57, spectrum 4), the peak splitting at 1550 cm^{-1} and the appearance of a broad band at 1800 cm^{-1} confirmed the location of Ag ions at the grain boundaries of DND nanoparticles [109]. Treatment in NH_3 plasma of the polymers (Fig. 58) revealed that the surface functional groups of the composites were significantly modified. This was proved by the decrease in the intensity of the bands for methyl groups, broadening and intensification of the bands in the region from 3000 to 3440 cm^{-1}. Thus, FTIR spectra confirmed that the composite layers treated with ammonia yielded increased hydrophilic properties.

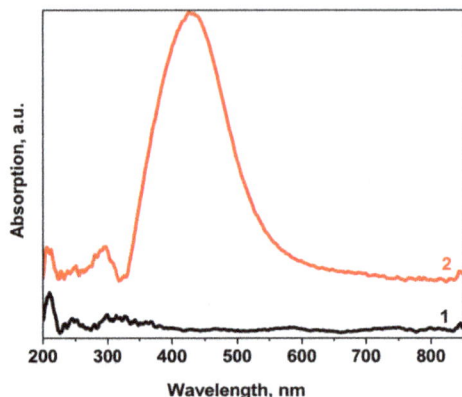

Figure 60. UV-visible absorption spectra of the synthesized DNDs: (1) DND; (2) Ag-DND.

The FTIR and UV-visible absorption spectra of the DNDs used in this work are presented in Figs. 59 and 60. The spectra provided evidence that all materials had hydrophilic surface which was chemically multifunctional. In the FTIR spectrum of the DND particles (Fig. 59, spectrum 1), the broad peak at 400 - 700 cm^{-1} was due to amorphous sp^3 bonded carbon, while the relatively sharp peaks at 1000 - 1150 cm^{-1} showed the presence of sp^2 bonded carbon atoms. The absorption peaks in the range 1500 - 1800 cm^{-1} were assigned to C=O stretching vibrations of different nature. The silanization of the DND surface (spectrum 2) was shown through the increase of the sp^2 bonded carbon atoms (peaks at 1000 - 1150 cm^{-1}) and the diminishing of the acidic surface functional groups observed in the region from 1500 to 1750 cm^{-1} due to formation of new Si-O-C bonds. The incorporation of Ag cations on the DND surface (spectrum 3) was revealed by the diminishing of the IR peaks in the region centered at 1100 cm^{-1}, and the appearance of new peaks at around 1300 cm^{-1}. The low intensive peaks in Fig. 60 at about 210 and 280 nm point to the formation of the Ag_2^+ and small silver clusters (Ag_n). The appearance of plasmon peak higher than 410 nm indicates the formation of Ag nanostructures [110,111].

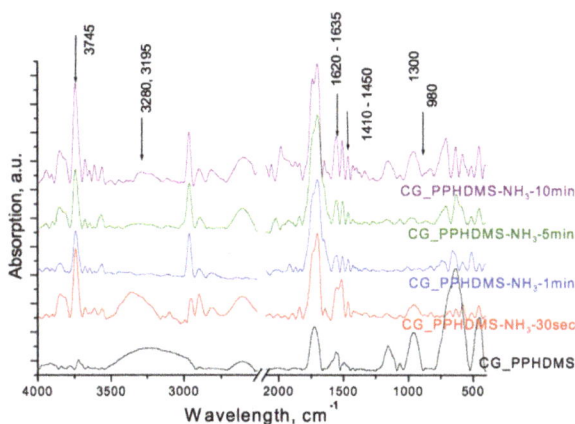

Figure 61. FTIR spectra of PPHMDS films, deposited on CG substrates: untreated (black) and treated with ammonia plasma for 30 s (red), 1 min (blue) and 10 min (violet).

The FTIR spectra after ammonia plasma modification for different treatment times (Fig. 61) demonstrated significant changes in the surface functionality. The comparison between the spectrum of untreated PPHMDS (black line) and the spectrum of PPHMDS treated with ammonia for 30 sec (red line) revealed the presence of NH^{4+} cations (new bands in the regions of 1410-1450 cm^{-1}; and 1620-1635 cm^{-1}) and NH_3 molecules (new bands at around 3745, 3280, 3195 and 990 cm^{-1}). Interaction occurred between the ammonia groups obtained in the plasma with Si-OH and OH$^-$ groups in the polymer structure as seen from the FTIR spectra in the characteristic for the polymer chain range 1200-800 cm^{-1}. With increasing the time of ammonia plasma treatment from 1 to 10 min, ammonia molecules were grafted onto the surface of PPHMDS until filling of all available polymer centers was achieved. A small shift and intensification of the bands around 1320, 1410-1450, 1630 and 1800 cm^{-1} was observed (Fig. 61, blue, olive and violet lines). Another notable change was the strong decrease in intensity of the bands characteristic of methyl groups (around 600 cm^{-1}) as well as broadening and intensification of the bands around 3750 cm^{-1} that determined the amount of grafted ammonia. Due to the reduction of the relative part of alkyl groups in the ammonia treated polymer, a decrease of the hydrophobic and an increase of the hydrophilic character of the surface were obtained. It is worth to note that in parallel experiments using KBr as substrates instead of CG only slight differences were observed in the FTIR spectra [112].

This observation confirms the importance of the CG substrate for the properties of plasma-polymerized layer and its surface functionality. It also shows that by ammonia plasma modification, layers with novel functional groups and more important, with strong hydrophilic properties could be obtained.

3.2   Biocompatibility Studies

Human osteoblast-like cells MG63 were maintained in Dulbeccos modified Eagle medium (DMEM, Sigma) supplemented with 10% FBS, 1% antibiotic/antimicotic mixture, 2 mM L-glutamine and 1 mM sodium pyruvate in a humidified atmosphere of 5% CO_2 in air. For cell experiments the MG63 cells were harvested with Trypsin/EDTA, subsequently washed, suspended in DMEM and seeded onto DNDs/PPHMDS films, individually placed in 24-well tissue culture plates at a density of 50 000 cells/well. Before cell seeding, half of the samples with composite films were pre-adsorbed with FN (Roche, Germany, 20 µg/ml solution in PBS) for 30 min at room temperature. The cells were incubated for 2 hours in serum-free culture medium DMEM before being stained for actin.

FDA staining was used to visualize overall cell morphology. For this purpose 2.0 x 10^4 cells were seeded onto each material in 1.0 ml medium without serum. One set of samples had been pre-coated with FN (20µg/ml) for 30 min at 37°C. After 4 hours of incubation, cells were stained with 0,001% FDA dissolved in acetone, incubated 2-3 min, rinse several times and then representative pictures of the adhered cells were taken using the green channel of a fluorescent microscope.

Q blue assay. For evaluation of cell proliferation a commercially available homogenous assay based on the conversion of the non-fluorescent dye resazurin into a high fluorescent product resorufin by metabolically active cells was used. For this purpose, adherent MG63 cells were incubated with Q Blue reagent at room temperature for 2 hours. At the end of incubation 100 µl from each sample was transferred into black 96-well plates in triplicates and the fluorescence intensity of the product is then quantified on a fluorescent microplate reader at 544 nm extinction and 590 nm emission filter. The results are expressed as mean ± standard deviation. Student's t-test was used for statistical analysis.

Alkaline phosphatase (ALP) activity was determined as 1 volume of cell lysate was incubated with 2 volumes of 0,2 mg/ml pNPP prepared in 2-Amino-2-methyl-1-propanol (AMP) buffer. In our experiments, 50 µl cell lysate was incubated with 100 µl pNPP-AMP in 96-well plates in triplicates. The enzyme reaction was allowed to proceed for 1,5 hours at room temperature until the yellow color appeared. Absorbance was read in a spectrophotometer at 405 nm. ALP activity is expressed as pmol pnitrophenol formed/min/g DNA or mg protein.

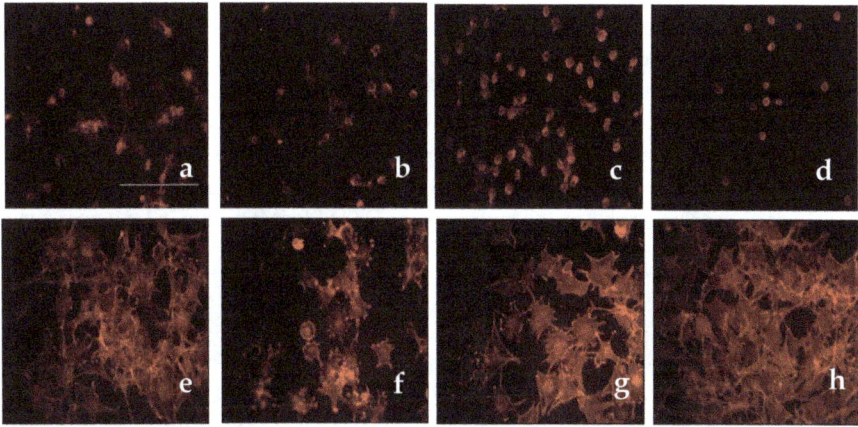

Figure 62. Overall morphology of MG63, cultured for 2 hours on plain CG slips, used as control (a,e), DND/PPHMDS (b,f); Si-DND/PPHMDS (c,g) and Ag-DND/ PPHMDS (d,h): non-coated (upper panel) and FN pre-coated (lower panel); bar 100μm.

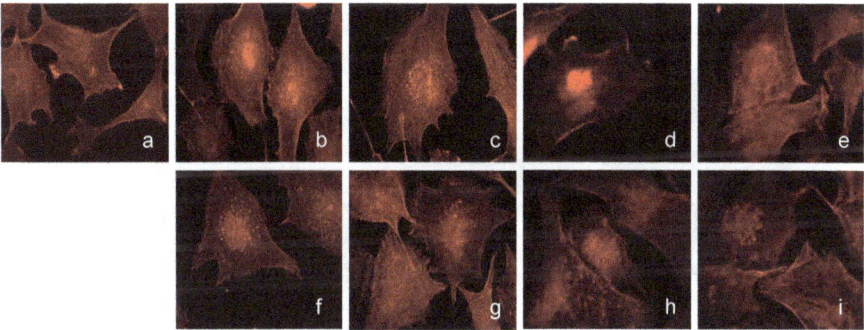

Figure 63. Actin cytoskeleton in MG63 osteoblast-like cells, cultured for 2 hours on FN pre-adsorbed CG slips used as control (a), PPHMDS (b,f), DND/PPHMDS (c,g), Ag-DND/PPHMDS (d,h) and Si-DND/PPHMDS (e,i): non-modified (upper panel) and NH₃-treated (lower panel).

To investigate the ability of the different polymer composites to support initial adhesion of osteoblast-like MG63 cells we examined overall cell morphology as well as intracellular component actin whose organization and structure change as a function of cell adhesion onto composite surfaces. Both stainings were compared for cells grown on glass. In addition, we studied effect of FN, which is the major adhesive protein in biological fluids. A change in the overall cell morphology and in the number ofthe attached cells in composites was demonstrated in Fig. 62. As expected, FN pre-coating additionally modulate the effect of DNDs/PPHMDS surface. On non-coated surfaces, the number of attached cells was the highest on Si-DND/PPHMDS (Fig. 62-c, upper panel), whereas on FN pre-coated surfaces cells adhered more readily on Ag-DND/PPHMDS (Fig. 62 d, low panel).

In general, FN improved cells adhesion properties of composite surfaces as indicated by the increased number of the attached and spread cells. On FN pre-coated DND/PPHMDS layers (Fig.62-b, low panel), cells with different morphology were distinguished, revealing different stages of cell adhesion. On Ag-DND/PPHMDS the osteoblast were well spread and exhibited organized cytoskeleton with well assembled actin stress fibers (Fig. 62, low panel and Fig. 63 d). In contrast, cells cultured on plain (FN non-coated) surfaces were small and round in shape (Fig. 62, upper panel) and no actin expression was seen.

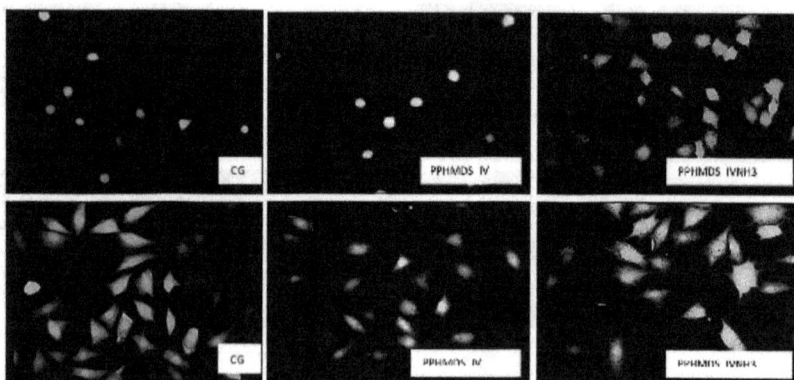

Figure 64. Overall morphology of MG63 osteoblasts, adhering for 2 hours on plain CG control and PPHMDS layers (upper panel) and FN-coated CG control and PPHMDS layers (lower panel); magnification 20x.

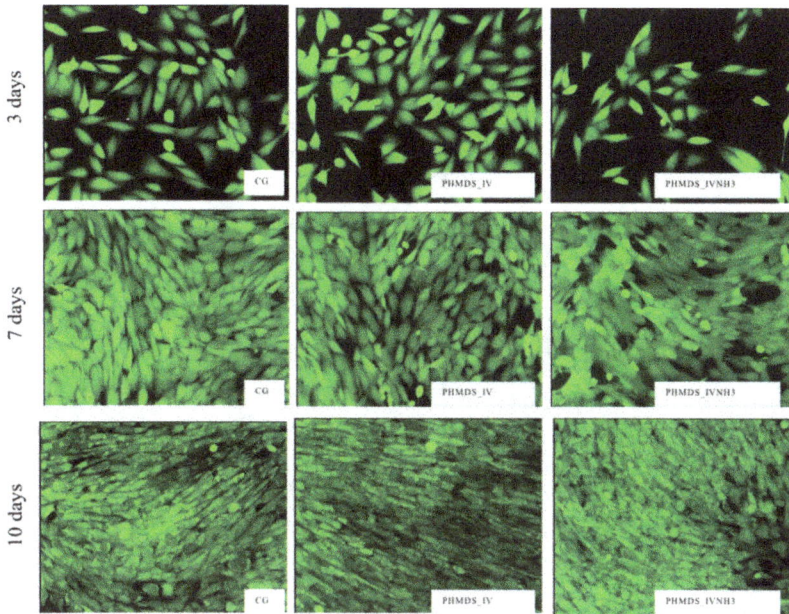

Figure 65. Overall morphology of MG63 osteoblasts, cultured for 3 days (upper panel), 7 days (middle panel) and 10 days (lower panel) on plain PPHMDS polymer layers and CG control; magnification 20x.

Further, we studied the effect of the NH_3 treatment of the PPHMDS layers on the cell morphology and adhesion. Overall morphology (Fig. 64) revealed that MG63 osteoblasts adhering for 2 hours attached poorly on unmodified PPHMDS films while hydrophylization enhanced both the amount of attached cells and their main spreading area. Precoating coating with FN improved significantly osteoblast adhesion but diminished the differences between samples. Nevertheless, cells spread slightly better on the NH_3 plasma treated PPHMDS layers.

In long-term cultures (cells cultivated for 3, 7 and 10 days), no significant differences among materials in terms of the cell viability and morphology were found (Fig. 65). In terms of cell growth however, NH_3-treated layers demonstrated significantly higher rate of cell proliferation compared to the hydrophobic PPHMDS (Fig. 66). ALP measurements demonstrated an increase on the PPHMDS layers at the day 10th (Fig. 67). In this case ammonia plasma deposition did not affect ALP activity of osteoblast cells.

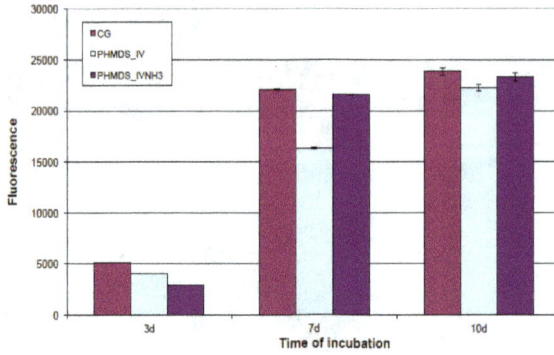

Figure 66. Cell proliferation of MG63 osteoblasts, cultured for 3, 7 and 10 days on PPHMDS layers, modified with NH₃ plasma and non-modified, and on control CG.

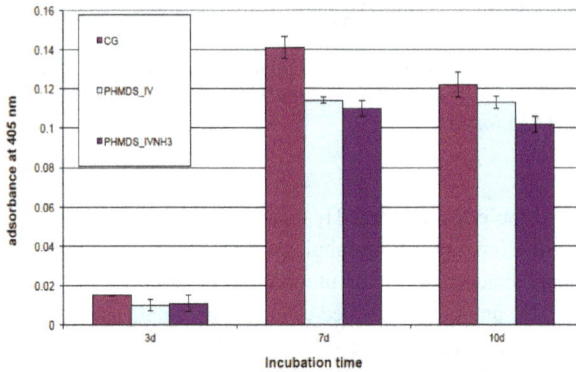

Figure 67. Phosphatase activity of MG63 osteoblasts, cultured for 3, 7 and 10 days on PPHMDS layers, modified with NH₃ plasma and non-modified, and on control CG.

3.3 Ability of ND Particles to Influence the PPHMDS Properties – Summary

In summary, our results suggest that ammonia plasma polymerization could improve interaction of osteoblasts with PPHMDS polymers and therefore has the potential to be applied in the field of bone tissue engineering.

The different cell morphology, cytoskeleton organization and focal adhesion formation were believed to be a result of the surface properties of the DNDs/PPHMDS composites that depend on the nanoparticles used as a polymer filler (DND, Si-DND or Ag-DND). Thus, not only the chemistry but the wettability and nanotopography were the factors that were changed in the different composites. It is known that the surface properties of biomaterials influence cell behavior through the adsorption characteristics of the serum proteins and proteins secreted by cells themselves [113]. It is not only the adsorbed quantity, but also protein conformation that will determine future cell behavior. FN pre-coating improved the cellular interaction to all layers and the cells adhered more readily especially on the Ag-DND/PPHMDS composite. By varying the DNDs type and their concentration it is possible to increase bioactivity and biocompatibility of the PPHMDS layer. These results encourage further studies on plasma polymer nanocomposites on various substrates for tissue engineering of scaffolds.

References

[1]   P. Calvert, S. Mann, Review: Synthetic and biological composites formed by in situ precipitation, J. Mater. Sci. 23 (1988) 3801-3806. https://doi.org/10.1007/BF01106796

[2]   L. Addadi, S. Weiner, Interactions between acidic proteins and crystals: Stereochemical requirements in biomineralization, Proc. Natl. Acad. Sci. USA 82 (1985) 4110-4114. https://doi.org/10.1073/pnas.82.12.4110

[3]   H. Lowenstam, S. Weiner, On Biomineralization, 1989, Oxford University Press, Oxford, pp 74-175.

[4]   A. Mount, A. Wheeler, R. Paradkar, D. Snider, Hemocyte-mediated shell mineralization in the eastern oyster, Science 304 (2004) 297-300. https://doi.org/10.1126/science.1090506

[5]   M.T. Pham, M.F. Maitz, H. Reuther, A. Muecklich, F. Prokert, G.J. Steiner, Biomed. Mater. Res. A 71 (2004) 16 https://doi.org/10.1002/jbm.a.30113

[6]   J. Hanker, B. Giammara, Biomaterials and biomedical devices, Science 242 (1988) 885-892. https://doi.org/10.1126/science.3055300

[7]   L. Gineste, M. Gineste, X. Ranz, A. Ellefterion, A. Guilhem, N. Rouquet, P. Frayssinet, Degradation of hydroxylapatite, fluorapatite, and fluorhydroxyapatite coatings of dental implants in dogs, J. Biomed. Mater. Res. 48 (1999) 224-234. https://doi.org/10.1002/(SICI)1097-4636(1999)48:3<224::AID-JBM5>3.0.CO;2-A

[8]   D. Pellenc, H. Berry, O.J. Gallet, Adsorption-induced fibronectin aggregation and fibrillogenesis, Colloid. Interface Sci. 298 (2006) 132-144. https://doi.org/10.1016/j.jcis.2005.11.059

[9]   Y. Zhai, F. Cui, Recombinant human-like collagen directed growth of hydroxyapatite nanocrystals, J. Cryst. Growth 291 (2006) 202-206. https://doi.org/10.1016/j.jcrysgro.2006.03.006

[10]   S. Chang, H. Chen, J. Liu, D. Wood, P. Bentley, B. Clarkson, Synthesis of a potentially bioactive, hydroxyapatite- nucleating molecule, Calcif. Tissue Int. 78 (2006) 55-61. https://doi.org/10.1007/s00223-005-0118-4

[11]   Y. Hu, C. Zhang, S. Zhang, Z. Xiong, J. Xu, Development of a porous poly(L-lactic acid)/hydroxyapatite/collagen scaffold as a BMP delivery system and its use in healing canine segmental bone defect, J. Biomed. Mater. Res. A 67 (2003) 591-598. https://doi.org/10.1002/jbm.a.10070

[12]   L. Baxter, V. Frauchiger, M. Textor, I. ap Gwynn, R. Richards, Fibroblast and osteoblast adhesion and morphology on calcium phosphate surfaces, Eur. Cells Mater. 4 (2002) 1-17. https://doi.org/10.22203/eCM.v004a01

[13]   H. Wan, R. Williams, P. Doherty, D.J. Williams, A study of cell behaviour on the surfaces of multifilament materials, Mater. Sci.: Mater. Med. 8 (1997) 45-51. https://doi.org/10.1023/A:1018542313236

[14]   J. Lee, H. Jung, I. Kang, H. Lee, Cell behaviour on polymer surfaces with different functional groups, Biomaterials 15 (1994) 705-711. https://doi.org/10.1016/0142-9612(94)90169-4

[15]   J. Schakenraad, H. Busscher, C. Wildevuur, J. Arends, The influence of substratum surface free energy on growth and spreading of human fibroblasts in the presence and absence of serum proteins, J. Biomed. Mater. Res. 20 (1986) 773-784. https://doi.org/10.1002/jbm.820200609

[16]   T. Horbett, J. Waldburger, B. Ratner, A. Hoffman, Cell adhesion to a series of hydrophilic-hydrophobic copolymers studied with a spinning disc apparatus, J. Biomed. Mater. Res. 22 (1988) 383-404. https://doi.org/10.1002/jbm.820220503

[17] M. Cote, C. Doillon, Wettability of Cross-Linked Collagenous Biomaterials: In Vitro Study, Biomaterials 1992, 13, 612-616. https://doi.org/10.1016/0142-9612(92)90029-N

[18] S. Mann, Dalton perspectives. Biomineralization: the form(id)able part of bioinorganic chemistry!, J. Chem. Soc., Dalton Trans. 1 (1993) 3953-3961.

[19] L. Suggs, M. Shire, C. Garcia, J. Anderson, A. Mikos, In vitro cytotoxicity and in vivo biocompatibility of poly(propylene fumarate-co-ethylene glycol) hydrogels, J. Biomed. Mater. Res. 46 (1999) 22-32. https://doi.org/10.1002/(SICI)1097-4636(199907)46:1<22::AID-JBM3>3.0.CO;2-R

[20] A. Jenney, J. Anderson, Effects of surface-coupled polyethylene oxide on human macrophage adhesion and foreign body giant cell formation in vitro, J. Biomed. Mater. Res. 44 (1999) 206-216. https://doi.org/10.1002/(SICI)1097-4636(199902)44:2<206::AID-JBM11>3.0.CO;2-D

[21] W. Chuang, T. Young, C. Yao, W. Chiu, Properties of the poly(vinyl alcohol)/chitosan blend and its effect on the culture of fibroblast in vitro, Biomaterials 20 (1999) 1479-1487. https://doi.org/10.1016/S0142-9612(99)00054-X

[22] B. Kasemo, J. Lausmaa, Surface science aspects on inorganic biomaterials, CRC Crit. Rev. Biocompat. 2 (1986) 335.

[23] Y. Ikada, Surface modification of polymers for medical applications, Biomaterials 15 (1994) 725-736. https://doi.org/10.1016/0142-9612(94)90025-6

[24] A. Dekker, K. Reitsma, T. Beugeling, A. Bantjes, J. Feijen, W. Van Aken, Adhesion of endothelial cells and adsorption of serum proteins on gas plasma-treated polytetrafluoroethylene, Biomaterials 12 (1991) 130-138. https://doi.org/10.1016/0142-9612(91)90191-C

[25] J. Dewez, A. Doren, Y. Schneider, P. Rouxhet, Competitive adsorption of proteins: Key of the relationship between substratum surface properties and adhesion of epithelial cells, Biomaterials 20 (1999) 547-559. https://doi.org/10.1016/S0142-9612(98)00207-5

[26] J. Ricci, J. Spivak, N. Blumenthal, H. Alexander, in: J. Davies (Ed.), The bone and biomaterial interface, University of Toronto Press: Toronto, Ontario, Canada, 1991.

[27] D. Hercules, Surface analysis: an overview, Crit. Rev. Surf. Chem. 1 (1992) 243-277.

[28] B. Kasemo, J. Lausmaa, in: J. Davies (Ed.), The bone and biomaterial interface; University of Toronto Press: Toronto, Ontario, Canada, 1991.

[29] D. MacDonald, B. Markovic, M. Allen, P. Somasundaran, A. Boskey, Surface analysis of human plasma fibronectin adsorbed to commercially pure titanium materials, J. Biomed. Mater. Res. 41 (1998) 120-130. https://doi.org/10.1002/(SICI)1097-4636(199807)41:1<120::AID-JBM15>3.0.CO;2-R

[30] M.T. Pham, H. Reuther, M.F. Maitz, Native extracellular matrix coating on Ti surfaces, J. Biomed. Mater. Res. A 66 (2003) 310-316. https://doi.org/10.1002/jbm.a.10575

[31] J. Elliot, Structure and chemistry of apatites and other calcium orthophosphates. Amsterdam, Elsevier Science, 1994

[32] E. Pecheva, L. Pramatarova, M.F. Maitz, M.T. Pham, A. Kondyuirin, Kinetics of hydroxyapatite deposition on solid substrates modified by sequential dual implantation of Ca and P ions. Part I. FTIR and Raman spectroscopy study, Appl. Surf. Sci. 235 (2004) 176-181. https://doi.org/10.1016/j.apsusc.2004.05.174

[33] E. Pecheva, L. Pramatarova, M.F. Maitz, M.T. Pham, A. Kondyuirin, Kinetics of hydroxyapatite deposition on solid substrates modified by sequential dual implantation of Ca and P ions. Part II. Morphological, composition and structure study, Appl. Surf. Sci. 235 (2004) 170- 175. https://doi.org/10.1016/j.apsusc.2004.05.178

[34] E. Pecheva, L. Pramatarova, G. Altankov, Hydroxyapatite grown on a native extracelluar matrix: initial interactions with human fibroblasts, Langmuir 23 (18) (2007) 9386-9392. https://doi.org/10.1021/la700435c

[35] E. Danen, K.Yamada, Fibronectin, integrins, and growth control, J. Cell. Physiol. 189 (2001) 1 https://doi.org/10.1002/jcp.1137

[36] K. Anselme, P. Linez, M. Bigerelle, D. Le Maguer, A. Le Maguer, P. Hardouin, H. Hildebrand, A. Iost, J. Leroy, The relative influence of the topography and chemistry of TiAl6V4 surfaces on osteoblastic cell behaviour, Biomaterials 21 (2000) 1567. https://doi.org/10.1016/S0142-9612(00)00042-9

[37] S. Nakagawa, P. Pawelek, F. Grinnell, Extracellular matrix organization modulates fibroblast growth and growth factor responsiveness, Exp. Cell Res. 182 (1989) 572-582. https://doi.org/10.1016/0014-4827(89)90260-7

[38]   D. Mears, Materials and Orthopaedic Surgery, The Williams and Wilkins Company: Baltimore, 1972.

[39]   K. Burg, S. Porter, J. Kellam, Biomaterial developments for bone tissue engineering, Biomaterials 21 (2000) 2347-2359. https://doi.org/10.1016/S0142-9612(00)00102-2

[40]   R. Narayan, Nanostructured diamondlike carbon thin films for medical applications, Mater. Sci. Eng. C 25 (2005) 405-416. https://doi.org/10.1016/j.msec.2005.01.026

[41]   T. Akasaka, F.Watari, Y. Sato, K. Tohji, Apatite formation on carbon nanotubes, Mater. Sci. Eng. C 26 (2006) 675-678. https://doi.org/10.1016/j.msec.2005.03.009

[42]   R. Hauert, A review of modified DLC coatings for biological applications, Diamond Relat. Mater. 12 (2003) 583-589. https://doi.org/10.1016/S0925-9635(03)00081-5

[43]   Y. Chen, Y. Q. Zhang, T. H. Zhang, C. H. Gan, C. Y. Zheng, G. Yu, Carbon nanotube reinforced hydroxyapatite composite coatings produced through laser surface alloying, Carbon 44 (2006) 37-45. https://doi.org/10.1016/j.carbon.2005.07.011

[44]   J. Wilson, W. Kulisch (eds), Diamond Thin Films, Akademie Verlag, Berlin, 1996.

[45]   O. Goudouri, S. Stavrev, X. Chatzistavrou, N. Kantiranis, T. Zorba, P. Koidis, K. Paraskevopoulos, Thermal behavior of a Bioactive Glass/Nanodiamonds system, Proc. XXII Panhellenic Conf. Solid State Phys. Mater. Sci., 24–27 September 2006, Patra, Greece, pp 1.

[46]   O. Mykhaylyk, Y. Solonin, D. Batchelder, R. Brydson, Transformation of nanodiamond into carbon onions: A comparative study by high-resolution transmission electron microscopy, electron energy-loss spectroscopy, x-ray diffraction, small-angle x-ray scattering, and ultraviolet Raman spectroscopy, J. Appl. Phys. 97 (2005) 074302. https://doi.org/10.1063/1.1868054

[47]   E. A. Ekimov, E. L.Gromnitskaya, S. Gierlotka, W. Lojkowski, B. Palosz, A. Swiderska-Sroda, J. A. Kozubowski, A. M. Naletov, Mechanical behavior and microstructure of nanodiamond-based composite materials, J. Mater. Sci. Lett. 21 (2002) 1699-1702. https://doi.org/10.1023/A:1020889129195

[48]   S. Catledge, M. Fries, Y. Vohra, in: H. S. Nalwa (Ed.), Encyclopedia of Nanoscience and Nanotechnology, vol. 7, 2003, p. 741.

[49]   V. Khabashesku, J. Margrave, E. Barrera, Functionalized carbon nanotubes and nanodiamonds for engineering and biomedical applications, Diamond Relat. Mater. 14 (2005) 859-866. https://doi.org/10.1016/j.diamond.2004.11.006

[50]   S. Rossler, A. Sewing, M. Stolzel, R. Born, D. Scharnweber, M. Dard, H.Worch, Electrochemically assisted deposition of thin calcium phosphate coatings at near-physiological pH and temperature, J. Biomed. Mater. Res. A 64 (2003) 655-663. https://doi.org/10.1002/jbm.a.10330

[51]   I. Zhitomirsky, Cathodic electrodeposition of ceramic and organoceramic materials. Fundamental aspects, Adv. Colloid Interface Sci. 97 (2002) 279-317. https://doi.org/10.1016/S0001-8686(01)00068-9

[52]   K. Kilpadi, P-L. Chang, S. Bellis, Hydroxylapatite binds more serum protein, purified integrins and osteoblast precursor cells than pure titanium or steel, J. Biomed. Mater. Res. A 57 (2001) 258-267. https://doi.org/10.1002/1097-4636(200111)57:2<258::AID-JBM1166>3.0.CO;2-R

[53]   J. Xie, M. Baumann, L. McCabe, Osteoblasts respond to hydroxyapatite surfaces with immediate changes in gene expression, J. Biomed. Mater. Res. A 71 (2004) 108-117. https://doi.org/10.1002/jbm.a.30140

[54]   L. Baxter, V. Frauchiger, M. Textor, I ap. Gwynn, R. Richards, Fibroblast and osteoblast adhesion and morphology on calcium phosphate surfaces, Eur. Cells Mater. 4 (2002) 1-17. https://doi.org/10.22203/eCM.v004a01

[55]   S. Stavrev, J. Karadjov, L. Markov, in: A. Proykova (Ed.), Meetings in Physics at University of Sofia, Heron Press: Sofia, 2001, pp 7.

[56]   R.Z. LeGeros, Properties of osteoconductive biomaterials: calcium phosphates, Clin. Orthop. Relat. Res. 395 (2002) 81-98. https://doi.org/10.1097/00003086-200202000-00009

[57]   M. Pourbaix, Atlas of Electrochemical Equilibria in Aqueous Solution, National Association of Chemical Engineers, Houston, TX, 1974.

[58]   L. Socrates, Infrared Characteristic Group Frequencies, Wiley, New York, 1980.

[59]   S. Leeuwenburgh, J. Wolke, J. Schoonman, J. Jansen, Influence of precursor solution parameters on chemical properties of calcium phosphate coatings prepared using Electrostatic Spray Deposition (ESD), Biomaterials 25 (2004) 641-649. https://doi.org/10.1016/S0142-9612(03)00575-1

[60] X. Wang, Y. Li, J. Wei, K. de Groot, Development of biomimetic nano-hydroxyapatite/poly(hexamethylene adipamide) composites, Biomaterials 23 (2002) 4787-4791. https://doi.org/10.1016/S0142-9612(02)00229-6

[61] C. Wagner, W. Riggs, L. Davis, J. Moulder, G. Muilenberg, Handbook of X-ray photoelectron spectroscopy, Perkin-Elmer Corporation, Physical Electronics Division, Eden Prairie, Minnesota, 1979.

[62] K. Anselme, Osteoblast adhesion on biomaterials, Biomaterials 21 (2000) 667-681. https://doi.org/10.1016/S0142-9612(99)00242-2

[63] R.G. Lebaron, K.A. Athanasiou, Extracellular matrix cell adhesion peptides: functional applications in orthopedic materials, Tissue Eng. 6 (2000) 85-103. https://doi.org/10.1089/107632700320720

[64] C. Reyes, T. Petrie, A. García, Mixed extracellular matrix ligands synergistically modulate integrin adhesion and signaling, J. Cell Physiol. 217 (2008) 450-458. https://doi.org/10.1002/jcp.21512

[65] I.D. Campbell, Biochem. Studies of focal adhesion assembly, Soc. Trans. 36 (2008) 263-266.

[66] G. Altankov, Th. Groth, Fibronectin matrix formation by fibroblasts on surfaces varying in wettability, J. Biomater. Sci. Polym. Edn. 8 (1997) 299-310. https://doi.org/10.1163/156856296X00318

[67] Th. Groth., G. Altankov, in: R.M. Ottenbrite, I. Sunamoto (Eds.), Frontiers in Biomedical Polymer Applications, Technomics Publisher Inc., Lancaster-Basel, 1998.

[68] Th. Groth, G. Altankov, in: P. Harris, D. Chapman (Eds.), New Biomedical Materials, IOS Press, 1998.

[69] C. Barrias, M. Cristina, L. Martins, G. Almeida-Porada, M. Barbosa, P. Granja, The correlation between the adsorption of adhesive proteins and cell behaviour on hydroxyl-methyl mixed self-assembled monolayers, Biomaterials 30 (2009) 307-316. https://doi.org/10.1016/j.biomaterials.2008.09.048

[70] T. Velling, J. Risteli, K. Wennerberg, D. Mosher, S. Johansson, Polymerization of type I and III collagens is dependent on fibronectin and enhanced by integrins alpha 11 beta 1 and alpha 2 beta 1, J. Biol. Chem. 277 (2002) 37377-37381. https://doi.org/10.1074/jbc.M206286200

[71] D.A. Wahl, J.T. Czernuszka, Collagen-hydroxyapatite composites for hard tissue repair, Eur. Cell. Mater. 11 (2006) 43–56. https://doi.org/10.22203/eCM.v011a06

[72] W. Zheng, W. Zhang, X. Jiang, Biomimetic collagen nanofibrous materials for bone tissue engineering, Adv. Eng. Mater. 12 (2010) 451–466. https://doi.org/10.1002/adem.200980087

[73] S.K. Padmanabhan, F. Gervasa, A. Sannino, A. Licciulli, Preparation and characterization of collagen/hydroxyapatite microsphere composite scaffolds for bone regeneration, Key Eng. Mater. 587 (2014) 239–244.

[74] K.Y. Lee, D.J. Mooney, Hydrogels for tissue engineering, Chem. Rev. 101 (2001) 1869–1877. https://doi.org/10.1021/cr000108x

[75] X. Liu, P.X. Ma, Polymeric scaffolds for bone tissue engineering, Ann. Biomed. Eng. 32 (2004) 477–486. https://doi.org/10.1023/B:ABME.0000017544.36001.8e

[76] F.F. Borghi, A.E. Rider, S. Kumar, Z.J. Han, D. Haylock, K. Ostrikov, Emerging stem cell controls: nanomaterials and plasma effects, J. Nanomater. 13 (2013) 1-15. https://doi.org/10.1155/2013/329139

[77] K. Saha, J.F. Pollock, D.V. Schaffer, K.E. Healy, Designing synthetic materials to control stem cell phenotype, Curr. Opin. Chem. Biol. 11 (2007) 381-387. https://doi.org/10.1016/j.cbpa.2007.05.030

[78] A. Colas, J. Curtis, Silicone biomaterials: history and chemistry & medical applications of silicones, in: B. Rathner, A. Hoffman, F. Schoen, J. Lemons (Eds.), Biomaterials Science: An Introduction to Materials in Medicine, second ed., Elsevier, Inc., New York, 2004.

[79] F. Briquet, A. Colas, X. Thomas, Silicones for Medical Use, Dow Corning France - European Healthcare Centre, Presented at the XIIIth "Technological Congress" "Polymers for Biomedical Use" Le Mans, March, France, 1996.

[80] E. Kulaga, L. Ploux, L. Balan, G. Schrofj, V. Roucoules, Mechanically responsive antibacterial plasma polymer coatings for textile biomaterials, Plasma Process. Polym. 1 (2014) 63-79. https://doi.org/10.1002/ppap.201300091

[81] Min-Hsien, Simple poly(dimethylsiloxane) surface modification to control cell adhesion, Surf. Interface Anal. 41 (2009) 11–16 https://doi.org/10.1002/sia.2964

[82] H. Muguruma, I. Karube, Plasma-polymerized films for biosensors, Trends in Anal. Chem. 18 (1999) 62-68. https://doi.org/10.1016/S0165-9936(98)00098-3

[83] A. Hiratsuka, H. Muguruma, R. Nagata, R. Nakamura, K. Sato, S. Uchiyama, I. Karube, Mass transport behavior of electrochemical species through plasma-polymerized thin film on platinum electrode, J. Memb. Sci. 175 (2000) 25-34. https://doi.org/10.1016/S0376-7388(00)00403-8

[84]   X. P. Zou, E. T. Kang, K. G. Neoh, Plasma-induced graft polymerization of poly(ethylene glycol) methyl ether methacrylate on poly (tetrafluoroethylene) films for reduction in protein adsorption, Surf. Coat. Technol. 149 (2002) 119-128 https://doi.org/10.1016/S0257-8972(01)01490-6

[85]   M. Loughran, S.-W. Tsai, K. Yokoyama, I. Karube, Simultaneous iso-electric focusing of proteins in a micro-fabricated capillary coated with hydrophobic and hydrophilic plasma polymerized films, Curr. Appl. Physics 3 (2003) 495-499. https://doi.org/10.1016/j.cap.2003.09.002

[86]   Y. Kase, H. Muguruma, Amperometric glucose biosensor based on mediated electron transfer between immobilized glucose oxidase and plasma-polymerized thin film of dimethylaminomethylferrocene on sputtered gold electrode, Anal. Sci. 20 (2004) 1143-1146. https://doi.org/10.2116/analsci.20.1143

[87]   H. Wang, D. Li, Z. Wu, G. Shen, R. Yu, A reusable piezo- immunosensor with amplified sensitivity for ceruloplasmin-based on plasma-polymerized film, Talanta 62 (2004) 201-208. https://doi.org/10.1016/S0039-9140(03)00424-7

[88]   A. Thalhammer, R.J. Edgington, L.A. Cingolani, R. Schoepfer, R.B. Jackman, The use of nanodiamond monolayer coatings to promote the formation of functional neuronal networks, Biomaterials 31 (2010) 2097–2104. https://doi.org/10.1016/j.biomaterials.2009.11.109

[89]   H. Huang, E. Pierstorff, E. Osawa, D. Ho, Active nanodiamond hydrogels for chemo-therapeutic delivery, Nano Lett. 7 (2007) 3305–3314. https://doi.org/10.1021/nl071521o

[90]   L.P. McGuinness, Y. Yan, A. Stacey, D.A. Simpson, L.T. Hall, D. Maclaurin, S. Prawer, P. Mulvaney, J. Wrachtrup, F. Caruso, R.E. Scholten, L.C.L. Hollenberg, Quantum measurement and orientation tracking of fluorescent nanodiamonds inside living cells, Nat. Nanotechnol. 6 (2011) 358–363. https://doi.org/10.1038/nnano.2011.64

[91]   Q. Zhang, V. Mochalin, I. Neitzel, I. Knoke, J. Han, C. Klug, J. Zhou, I. Lelkes, Y. Gogotsi, Mechanical properties and biomineralization of multifunctional nanodiamond-PLLA composites for bone tissue engineering, Biomaterials 32 (2011) 87–94. https://doi.org/10.1016/j.biomaterials.2010.08.090

[92]   L. Grausova, L. Bacakova, A. Kromka, S. Potocky, M. Vanecek, M. Nesladek, V. Lisa, Nanodiamond as promising material for bone tissue engineering, J. Nanosci. Nanotechnol. 9 (2009) 3524–3534. https://doi.org/10.1166/jnn.2009.NS26

[93] K. Hristova, E. Pecheva, L. Pramatarova, G. Altankov, Improved interactions of osteoblast-like cells with apatite-nanodiamond coatings depends on fibronectin, J. Mater. Sci. Mater. Med. 22 (8) (2010) 1891–1900. https://doi.org/10.1007/s10856-011-4357-9

[94] E. Pecheva, L. Pramatarova, T. Hikov, Y. Tanaka, H. Sakamoto, H. Doi, Y. Tsutsumi, T. Hanawa, Apatite-nanodiamond composite as a functional coating of stainless steel, Surf. Interface Anal. 42 (2010) 475–480. https://doi.org/10.1002/sia.3213

[95] V. Mochalin, O. Shenderova, D. Ho, Y. Gogotsi, The properties and applications of nanodiamonds, Nat. Nanotechnol. 7 (2012) 11–23. https://doi.org/10.1038/nnano.2011.209

[96] A.M. Schrand, H. Huang, C. Carlson, J.J. Schlager, E. Osawa, S.M. Hussain, L. Dai, Are diamond nanoparticles cytotoxic? J. Phys. Chem. B 1 (2007) 2–7. https://doi.org/10.1021/jp066387v

[97] N. Mohan, C.-S. Chen, H.-H. Hsieh, Y.-C. Wu, H.-C. Chang, In vivo imaging and toxicity assessments of fluorescent nanodiamonds in Caenorhabditis elegans, Nano Lett. 10 (2010) 3692–3699. https://doi.org/10.1021/nl1021909

[98] V.A. Popov, Metal matrix composites with non-agglomerated nanodiamond reinforcing particles, in: X. Wang (Ed.), Nanocomposites, Nova Science Publishers, Inc., USA, 2013, pp. 369–401.

[99] S.A. Rakha, R. Raza, A. Munir, Reinforcement effect of nanodiamond on properties of epoxy matrix, Polym. Compos. 34 (2013) 811–818. https://doi.org/10.1002/pc.22480

[100] L. Pramatarova, E. Radeva, E. Pecheva, T. Hikov, N. Krasteva, R. Dimitrova, D. Mitev, P. Mongomery, R. Sammons, G. Altankov, The advantages of polymer composites with detonation nanodiamond particles for medical applications, in: L. Pramatarova (Ed.), Biomimetics, InTech Inc., Croatia, 2011, pp. 297–320. https://doi.org/10.5772/22903

[101] G. Schmidt, M.M. Malwitz, Properties of polymer–nanoparticle composites, Curr. Opin. Colloid Interface Sci. 8 (2003) 103–108. https://doi.org/10.1016/S1359-0294(03)00008-6

[102] J. Wei, M. Yoshinary, S. Takemoto, M. Hattori, E. Kawada, B. Liu and Y. Oda, J Boimed. Mater. Res. B 81 (2007) 66-75. https://doi.org/10.1002/jbm.b.30638

[103] N. Inagaki, Y. W. Park, K. Narushim, K. Miyazaki, Plasma modification of poly(oxybenzoate-co-oxynaphthoate) film surfaces for copper metallization, J. Adhesion Sc. Technol. 18 (2004) 1427-1447. https://doi.org/10.1163/1568561042323275

[104] D. Mitev, R. Dimitrova, M. Spassova, C. Minchev, S. Stavrev, Surface peculiarities of detonation nanodiamonds in dependence of fabrication and purification methods, Diamond Relat. Mater. 16 (2007) 776–780. https://doi.org/10.1016/j.diamond.2007.01.005

[105] K. Vasilev, V. Sah, K. Anselme, C. Ndi, M. Mateescu, B. Dollmann, P. Martinek, L. Ploux, H. Griesser, Tunable antibacterial coatings that support mammalian cell growth, Nano Letter. 10 (2010) 202-207. https://doi.org/10.1021/nl903274q

[106] A. Agarwal, T. Weis, M. Schurr, N. Faith, C. Czuprynski, J. McAnulty, Ch. Murphy, N. Abbott, Surfaces modified with nanometer-thick silver-impregnated polymeric films that kill bacteria but support growth of mammalian cells, Biomaterials 31 (2010) 680-690 https://doi.org/10.1016/j.biomaterials.2009.09.092

[107] N. Benissad, K. Aumaille, A. Granier, A. Goullet, Structure and properties of silicon oxide films deposited in a dual microwave-rf plasma reactor, Thin Solid Films 384 (2001) 230-235. https://doi.org/10.1016/S0040-6090(00)01870-8

[108] R. A. B. Devine, Densification-induced infrared and Raman spectra variations of amorphous SiO2, J. Vac. Sci. Technol. A 6 (1988) 3154-3156. https://doi.org/10.1116/1.575047

[109] L. Pramatarova, N. Krasteva, E. Radeva, E. Pecheva, R. Dimitrova, T. Hikov, D. Mitev, K. Hristova, G. Altankov, Study of detonation nanodiamond – plasma polymerized hexamethildisiloxan composites for medical application, J. Phys. Conf. Ser. 253 (2010) 012078. https://doi.org/10.1088/1742-6596/253/1/012078

[110] C-S. Hong, H. Park, S. Wang, J. Moon, H. Park, R. Hill, Formation of photoresist-free patterned ZnO film containing nano-sized Ag by photochemical solution deposition, Appl. Surf. Sci. 252 (2006) 7739–7742. https://doi.org/10.1016/j.apsusc.2006.03.070

[111] J. Murphy, N.R. Jana, Controlling the aspect ratio of inorganic nanorods and nanowires, Adv. Mater. 14 (2002) 80–82 https://doi.org/10.1002/1521-4095(20020104)14:1<80::AID-ADMA80>3.0.CO;2-#

[112] N. Krasteva, G. Toromanov, K. Hristova, E. Radeva, E. Pecheva, R. Dimitrova, G. Altankov, L. Pramatarova, Initial biocompatibility of plasma polymerized hexamethyldisiloxane films with different wettability, J. Phys. Conf. Ser. 253 (2010) 012079 https://doi.org/10.1088/1742-6596/253/1/012079

[113] C. Barrias, M. Martins, G. Almeida-Porada, M. Barbosa, P. Granja, The correlation between the adsorption of adhesive proteins and cell behaviour on hydroxyl-methyl mixed self-assembled monolayers, Biomaterials 30 (2009) 307-316. https://doi.org/10.1016/j.biomaterials.2008.09.048

CHAPTER 5

Development of *In vitro* Bacterial Models of Dental Plaque

Abstract

Bacteria are some of the oldest living organisms on this planet. They were the dominant form of life on Earth for more than 3 billion years, and still form much of the world's biomass. Bacteria are closely associated with almost every aspect of human life. Bacterial infection is a major cause of human and animal diseases and even death. On the other hand, bacteria can be beneficial: for instance in biomineralization - a process of accumulation of various minerals by microbial organisms from the environment.

Any group of microorganisms sticking to each other and adhering on a surface is called a bacterial film (also known as biofilm). The microorganisms are embedded within a self-produced matrix of extracellular polymeric substances (including extracellular DNA, proteins and polysaccharides) [1]. They form bacterial communities, which adhere to any surface which is in contact with a non-sterile media in the environment or in our body, and are involved in a wide variety of microbial infections in the body. One of the most common biofilm communities is found in the oral cavity on teeth (known as *dental plaque*) and it is a tenacious adherent bacterial commune that contains potentially pathogenic bacteria. Dental plaque must be removed regularly to prevent the onset of caries, where tooth enamel is eroded from acids produced by the bacteria in the biofilm. The bacteria can then break through and invade the tubular structure of dentine leading into the soft tissue of the pulp. Periodontal diseases are a result of oral biofilm remaining on the surfaces of teeth and invoke destructive processes by the body that leads to bone loss, dental abscess formation and eventual tooth loss or other serious infections of the oral cavity. Dental biofilm can also mineralize into hard calculus, which is very difficult to remove with normal cleaning techniques such as brushing, flossing and mouthwash. This allows the biofilm to continue forming and exacerbating the resulting problems. Calculus can be removed by a dentist using hand instruments or an ultrasonic scaler. The latter is a preferred form of treatment modality by dental health professionals. However, these instruments are not 100% efficient, and in order to improve them, it is essential to study biofilm growth, removal and structure. Imaging techniques are integral for this work.

To assist in the study of biofilm formation and removal from confined spaces, flat or complex surfaces, a few types of bacteria were used to develop *in vitro* biofilm model systems on various model surfaces. Many naturally occurring mineralization processes yield HA or related salts, but biological routes to calcification have not generally been exploited for production of HA for clinical applications. The first model system that is presented here was built around *Serratia* NCIMB40259, which is a non-pathogenic gram-negative bacteria that can form a biofilm on various surfaces (glass, Ti, polymers, etc.) [2,3]. It can also enzymatically catalyze the *in situ* formation of HA around the bacterial cells attached to the surface under study and thus mimic the formation of calculus on teeth (calculus is the mineralized biofilm) [2-4]. Unlike the bacteria involved in calculus formation, *Serratia* biofilms are easy to grow and mineralization is readily achieved making it potentially an excellent model system for this application. However, because *Serratia* bacteria is not an oral bacterium biofilms were established with oral bacteria including *Streptococcus mutans* as representative of the most prevalent species in human plaque and strongly associated with caries production in teeth [5]. In terms of materials to grow the biofilms on, the following were chosen as models of internal or external teeth surfaces. A root canal model was developed to represent the confined spaces within the tooth (Figs. 68, 69). Flat glass and Ti surfaces were also used: glass is easy to be imaged and perform clear image analysis, and Ti is widely used as dental implant material due to its biocompatibility. A cheek model was developed to represent the bioflora on cheek and tongue which are responsible for bad mouth odor [6]. Finally, a tooth model with real teeth was developed to image real situations in the mouth where there are interdental spaces and gum.

Keywords

Bacteria, Oral Plaque, *In Vitro* Models

Contents

1. Growth of Bacteria Inside a Root Canal Model

A tooth can roughly be divided into crown and root(s) (Fig. 70). The roots contain a hollow space called the root canal, which is filled with pulp tissue (nerves, undifferentiated cells, connective tissue and blood vessels) that supplies the tooth with nutrients [7]. The root canal system is geometrically very complex, as the root canal can be curved and have side canals, apical ramifications and isthmuses [8]. Fig. 70 a shows an example of clinically encountered root canal geometry. The wall of the root canal is furthermore made of dentin and is a porous structure, as it contains microchannels (tubules) with a diameter of 0.5 to 3.2 µm with a density of 10^3-10^4 tubules/mm^2 [9]. In this work, the root canal system has been simplified to a straight, unbranched frustum of a cone (Fig. 70 b), in order to have a standardized root canal model that allows for understanding the influence of the geometry on the flow inside the root canal.

The *in vitro* root canal models were prepared from silicone rubber and consisted of a main canal and a perpendicular side canal situated on 3 mm distance from the apical of the main canal (Fig. 69). The length of the main canal was 20 mm, the apical (the narrowest end) and coronal (the widest end) diameters varied between 0.26-0.52 mm and 0.63-1.5 mm, respectively. The diameter and length of the side canal were 0.2 mm and 10 mm, respectively.

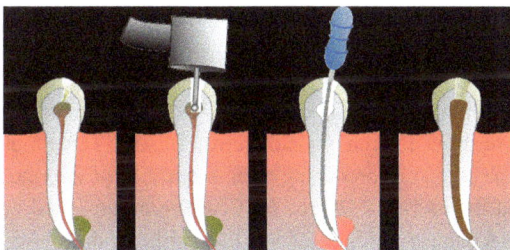

Figure 68. The root canal of the tooth and how dental instruments access it.

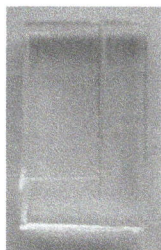

Figure 69. Silicone model of the root canal.

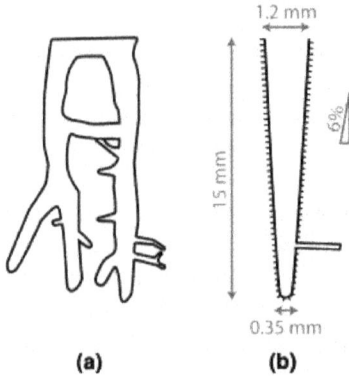

Figure 70. (a) Scheme of the root canal geometry as encountered clinically and (b) a simplified model with dimensions and side canal indicated.

Serratia NCIMB40259 was grown at 30°C on nutrient agar in aerobic conditions and a single colony was used to inoculate 50 ml of minimal essential medium (MEM: 12 g/l TRIS buffer adjusted to pH 7.3, 0.62 g/l KCl, 0.96 g/l $(NH_4)_2HPO_4$, 0.063 g/l $MgSO_4$-$7H_2O$, 0.00032 g/l $FeSO_4$-$7H_2O$, dissolved in distilled water) for 72 hours at 30°C and 150 rpm with air flow. The acid phosphatase (AP) enzyme specific activity of the bacteria culture is an important parameter showing the ability of the bacteria to form biofilm on the desired surface [10]. The optical density at 600 nm (OD_{600}) and AP activity of the planktonic bacteria produced in the primary bacterial culture were measured UV-Vis spectrophotometer (Jenway 7315, Bibby Scientific, Hong Kong). Since the AP of the primary culture was not sufficiently high (it was bellow 1000), 10 ml of the primary culture were used to inoculate 100 ml of lactose MEM (MEM with added lactose at a concentration of 0.6 g/l) to grow a secondary culture whose AP was high, at over 1000 units, showing that the bacteria were producing a large quantity of enzyme. The secondary culture was incubated for further 72 hours (30°C, 150 rpm) and used to inoculate the bioreactor. The root canal models were loaded into the bioreactor on holders under aseptic conditions and the assembled bioreactor (Fig. 71 a) was autoclaved at 120°C for 20 minutes at 1 bar pressure (Prestige Medical autoclave, UK). Further, the bioreactor was aseptically inoculated with 30 ml of the secondary *Serratia* culture which was left to grow batch-wise (i.e. no inflow of fresh medium) at 30°C while the bacteria were in their exponential phase of growth for 24 hours under aerobic conditions. The bioreactor was then switched to continuous mode, with fresh medium being pumped in continuously (approx. 34 ml/hour) while the bacteria were in their stationary phase of growth for 6 days until a biofilm layer was visible on the models. After the end of the biofilm growth, the silicone models were fixed in 2.5% glutaraldehyde buffer in 0.1 M sodium cacodylate buffer, pH 7.3. They were then dehydrated in a series of ethanol solutions (from 20 – 100%; 10 minutes each) followed by 100% hexamethyldisilasane [11]. Using fixation of the biofilms was a necessity because of the utilization of imaging techniques for the quantification of the biofilm removal, which required dried and fixed biofilms. SEM (Zeiss EVO MA-10, Cambridge, UK, 5 kV, 10 mm working distance) and Nikon digital camera were used for imaging, EDX spectroscopy for chemical composition and FTIR (Bomem IR spectrometer FTLA2000, ABB Inc., reflection mode, 4 cm^{-1} resolution) for the mineral identification.

Models with pre-grown biofilms were arranged in a glass flask filled with mineralization solution containing 50 mM AMPSO buffer ($C_7H_{16}NNaO_5S$, pH=8.6, Sigma, UK), 25 mM $CaCl_2$ (Sigma, UK) and 50 mM β-glycerol phosphate (β-GP, $C_3H_9NaO_6P$, Sigma, UK), based on a previous study results which demonstrated that this mixture gave the highest crystal yield [12]. The flask was fixed to an orbital shaker at 100 rpm and room temperature, and left for 24 hours for allowing nucleation. Then the mineralization medium was changed with a fresh one and the biofilm-coated models were left for five days. The models with mineralized biofilm were washed with distilled water and dried at 60°C for 24 hours (Fig. 71 b).

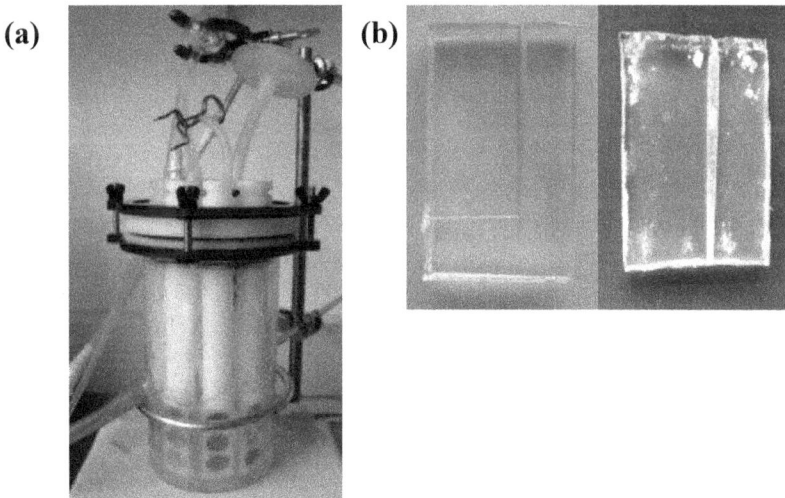

Figure 71. Set-up for the Serratia growth: (a) bioreactor used for the growth; (b) the silicone root canal model as it is (left) and after the biofilm mineralization (right).

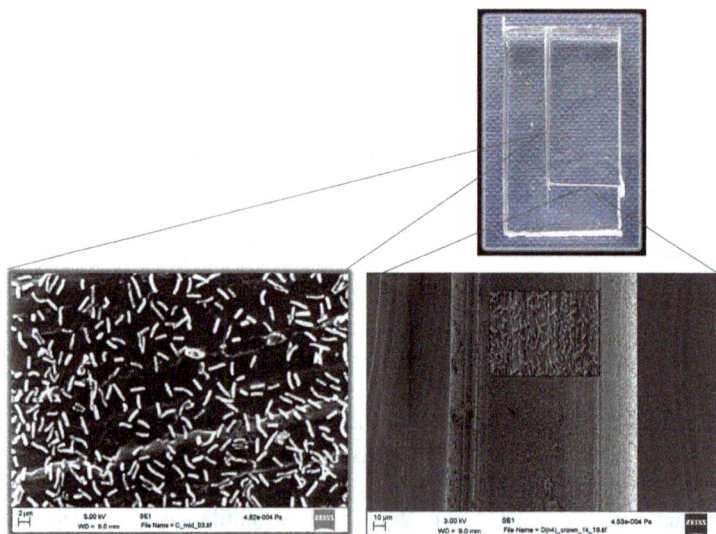

Figure 72. In vitro root canal model (image on the right). Second row: SEM images of Serratia bacteria grown along the main (left) and the side (right) canals; inset shows higher magnification of the bacteria within the side canal.

SEM study showed that the *Serratia* bacteria were equally present in the coronal part of the tooth canal model, as well as in the apical part, independent on the canal diameter (Fig. 72). Although the side canal was much narrower it was also occupied by the bacteria. The *Serratia* bacilli were approximately 2 μm in length. After the models have been left in the mineralization solution, the biofilm was covered by white deposits (Fig. 73). EDX spectroscopy coupled with the SEM images showed that the samples contain calcium, phosphorus and oxygen, therefore one or more CaP phases are present in the deposits. The high percentage of carbon shown in the samples may be due to the carbon in the organic content of the sample. Sodium and chloride were also evident in the organic phase and silicon came from the model. EDX is a useful tool for calculating the calcium to phosphorus ratio of CaPs. The results showed that the Ca:P ratio of the crystals produced in the bioreactor was 1.24 ± 0.02. From this information, the CaP phases present in the sample can be deduced: Ca:P ratio of 1.24 suggests that the deposits contain low crystalline carbonate-containing HA.

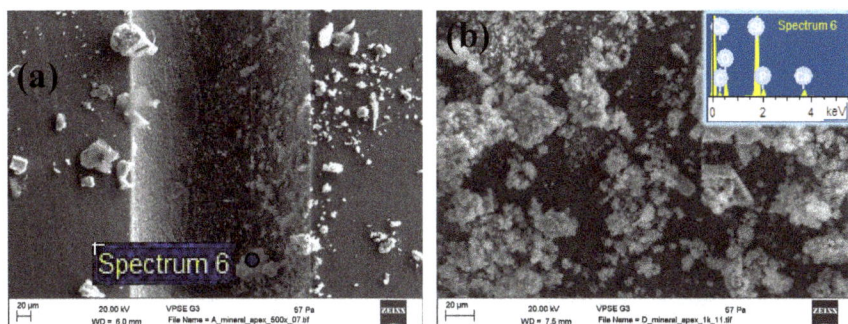

Figure 73. SEM image of: (a) the mineralized biofilm within the root canal (x500) and (b) higher magnification of (a) (x1000); inset: EDX spectrum of the mineralized biofilm.

FTIR spectroscopy revealed more details (Fig. 74) and confirmed that carbonated, Ca-deficient HA was formed in the root canals. The peaks at 550-1100 cm^{-1} were due to P-O vibrations in the phosphate (PO_4) group as follows: O-P-O deformation at 610-550 cm^{-1} and P-O asymmetric stretching at 960-970 and 1020-1100 cm^{-1}. Acid phosphate (HPO_4) has also contribution to the spectrum; two P-OH stretching vibrations at 880-950 and 2700-3400 cm^{-1}, as well as a P-O-H deformation at 1150-1280 cm^{-1} were observed. Carbonate was detected at 870-880 and 1400-1480 cm^{-1} and was assigned to the O-C-O deformation and C-O stretching vibrations. Water and hydrocarbon content were identified due to the H-O-H deformation at 1645-1655 cm^{-1} and C-H stretching at 2800-3000 cm^{-1}.

The results in this section showed that *Serratia sp* biofilms were able to grow inside the root canal model and mineralization was readily achieved making it a suitable model system for studying biofilm and calculus formation. Further development of the model will provide essential information for periodontal and endodontic treatment potentially leading to the development of a new range of instruments for cleaning inside root canals and other inaccessible places wherever biofilms are a problem.

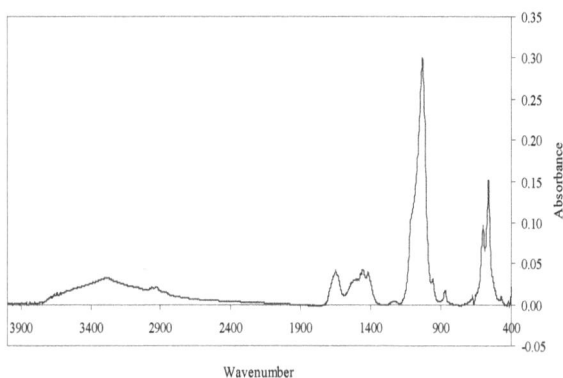

Figure 74. FTIR spectrum of the mineralized biofilm; peaks are characteristic of HA.

2. Growth of Bacteria on Flat Surfaces of Glass and Titanium

In vitro biofilms were grown on flat glass slides (20 x 10 x 1 mm in size) and Ti disks (grade 2, diameter 15 mm) in the bioreactor using *Serratia* NCIMB40259 or *Streptococcus mutans* ATCC 25175TM. Gram-positive *S. mutans* is one of the most prevalent species in human dental plaque, strongly associated with caries production in teeth and its removal is of concern in oral hygiene [13,14]. Reproducible bacterial biofilms (*Serratia* and *S. mutans*) with a thickness of about 2 μm were grown; this thickness was considered efficient for imaging the biofilm and its disruption. Glass is easy to be imaged and perform clear image analysis, which justified its choice as a model substrate. Ti was chosen because it is easier to image the biofilm removal from a flat metal substrate than from a real Ti implant or from the more frequently used HA disks due to the high roughness of the latter.

Serratia was grown on autoclaved glass and Ti (120°C, 20 min) following the procedure described above (section 1). To cultivate *S. mutans* biofilms, bacteria from a single colony growing on tryptone soya agar (Oxoid) were inoculated in 100 ml of tryptone soya broth (TSB, Oxoid) for 48 hours in an anaerobic chamber at 37°C and 100 rpm. The autoclaved substrates (glass or Ti) were aseptically fixed on the reactor rods and 30 ml of the inoculum was pipetted into the bioreactor, which was then topped up with TSB containing 0.5% sucrose. The bacteria were incubated without aeration but continuously stirred at 100 rpm in batch phase (without feeding) for 24 hours at 37°C to allow them to adhere to the discs before starting the continuous feeding mode, which was continued for 4 days to obtain a visible biofilm. Different durations for the biofilms growth were tried and a biofilm thickness of 2 μm was obtained with *Serratia* species for 6 days and with *S. mutans* for 4 days.

Serratia biofilms (Fig. 75) were imaged by LM (Nicon Eclipse TE300, Japan) and SEM (Zeiss EVO MA-10, Germany) and revealed the growth of a dense and homogeneous films on both glass and Ti substrates with *Serratia* bacilli having rod shape typical of the gram-negative species and length of approximately 2 μm. *S. mutans* biofilms (Fig. 76) on both materials were denser and forming clumps of bacteria better expressed on glass. The round *S. mutans* bacteria were organized in chains as seen on glass, which is characteristic of the gram-positive bacterial species.

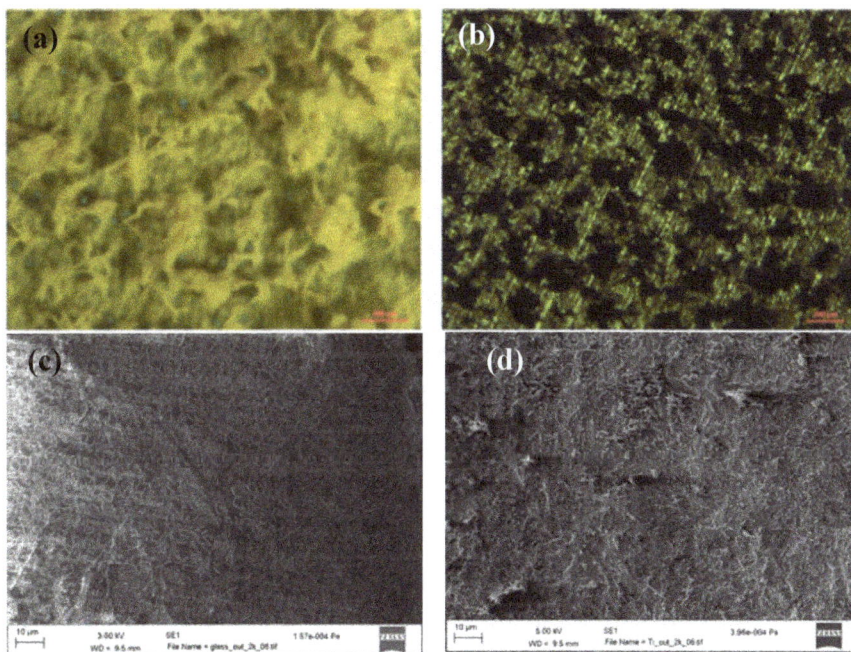

Figure 75. LM (first row, scale bar 200 μm) and SEM images (second row) of the Serratia biofilms grown on glass (a, c) and Ti (b, d).

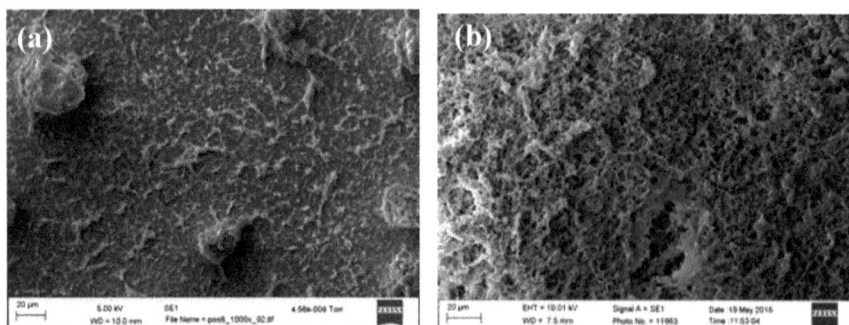

Figure 76. SEM images of the S. mutans biofilm grown on glass (a) and Ti (b).

3.   Growth of Bacteria on a Tooth Model

Tooth models were prepared with natural teeth using two lower molars (6th and 7th) and one lower premolar (5th), aligned and embedded in a pink poly(methyl methacrylate) (PMMA) acrylic matrix (Fig. 77 a) for growing *in vitro* plaque (Fig. 77 b). Biofilms of *S. mutans* were used as a model *in vitro* plaque in this work because it is the most prevalent species in human plaque and is strongly associated with caries production in teeth so its removal is of concern in oral hygiene [5].

The absorption of bacteria-containing culture medium at different periods of time (2 to 6 days) which gives an estimation how the number of bacterial cells was growing was measured by UV-Vis spectrophotometer. Nikon D40 digital camera was used to take general pictures of the tooth models. LM and SEM were used for imaging the tooth models with the grown biofilm in details.

A preliminary study of the kinetics of biofilm growth, i.e. how the bacterial film grows in thickness with time was performed. This was required in order to decide how many days are needed to grow *in vitro* plaque visible with a naked eye on the models due to their complex shape. Tooth models were prepared with thermoplastic teeth (Frasaco, Germany) for this study. Six tooth models were aseptically fixed in the bioreactor and the procedure described in section 2 was followed. Every day one model was taken out for counting the bacteria adhered on the teeth surface, as well as to optically estimate the thickness of the growing plaque. Every tooth model with *in vitro* plaque was rinsed three times by immersion in 30 ml of sterile PBS to remove loosely bound bacteria. The plaque was then manually brushed off with a dry sterile toothbrush (TB) using 20 linear strokes (forth and back) which lasted approximately 10 sec, and the TB was placed in an universal tube with 10 ml of PBS for 10 min. Subsequently the tube was sonicated for 30 sec to detach the bacteria adhered to the TB filaments. Serial dilutions were then prepared and 100 µl of each of them was spread with a sterile loop over two tryptone soya agar (TSA) plates per each dilution and incubated in an aerobic incubator at 37 °C for 48 hours [15]. Afterwards, the number of bacterial colonies was counted manually from the TSA plate and by measuring the absorbance with UV-Vis spectrophotometer at 620 nm (Table 9). It was concluded that an optimal *in vitro* plaque to perform the brushing experiments was grown at day 3 (Fig. 78). After day 3 there were no significant changes in the number of growing bacteria. The procedure for growing *S. mutans* biofilms described in section 2 was followed for growing the *in vitro* plaque on real tooth models.

Table 9. Counting the number of bacterial colonies

Day 1	Day 2	Day 3	Day 4	Day 5
CFU/ml (colony forming units) - counting bacteria after brushing [cells/ml]:				
1304	644	102	90	120
Spectrophotometry – absorption of bacteria at 620 nm [a.u.]:				
0.094	0.143	0.112	0.146	0.120

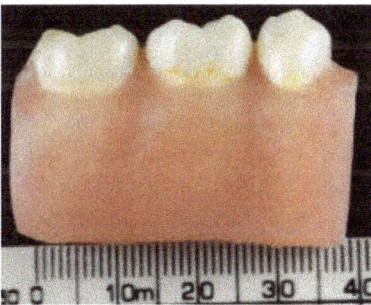

(a) natural teeth model

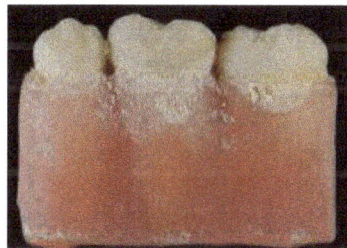

(b) teeth model with grown S. mutans plaque

Figure 77. Natural tooth model before (left) and after (right) in vitro plaque growth.

Day 3

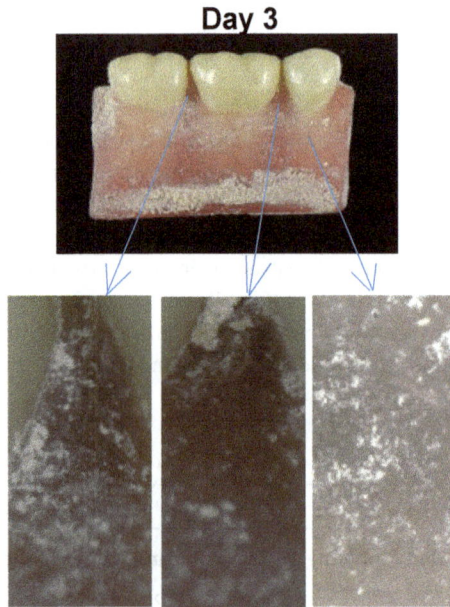

Figure 78. Images of the S. mutans biofilm grown on the 3rd day on the plastic tooth model showing in details how the biofilm has covered the two interdental spaces and the gum (arrows).

4.   Growth of Bacteria on a Cheek Model

Little information is available on the microbiota of mucosal surfaces. The oral mucosa of the gingiva, palate, cheeks, and floor of the mouth are colonized with few microorganisms [6]. Streptococci constitute the highest proportion of the microbiota in these sites, with a predominance of *S. Sanguinis* and *S. Oralis*. On the tongue, even higher bacterial density and diversity is found [6]. In all studies, *Streptococcus* sp. (*S. Salivarius* and *S. Mitis*) were the predominant members of the microbiota. Thus, we have chosen to grow mixed biofilm of *S. Sanguinis* and *S. Mutans* as a model *in vitro* biofilm on the cheek models.

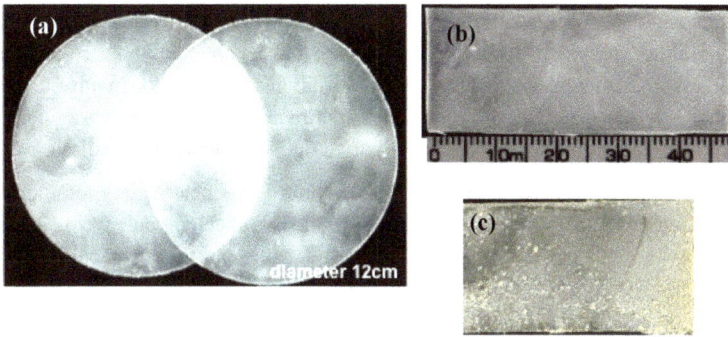

Figure 79. Cheek model preparation: (a) sand-blasted vinyl sheets for cutting the model; (b) vinyl strip cut from the spherical sheet (c) the strip with grown in vitro biofilm.

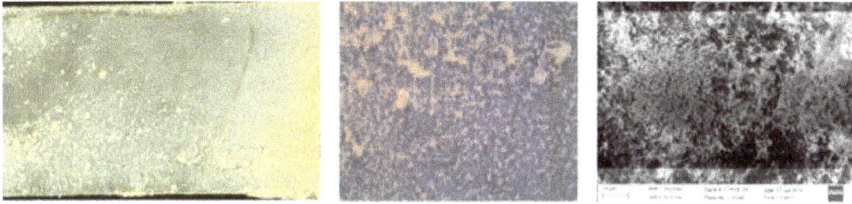

Figure 80. Images of the cheek model with grown two-species biofilm, taken with a digital camera (left), LM (middle) and SEM (right).

Cheek models were prepared from sandblasted spherical vinyl sheets (Fig. 79 a), usually used for biting prosthesis in dentistry, which makes it biocompatible, as well as because its elasticity resembles the one of cheeks. Sandblasting was performed to increase the surface roughness and thus allowing the bacteria to adhere better. Strips were cut in 20 x 50 mm size (Fig. 79 b) for growing *in vitro* biofilm (Fig. 79 c).

A single colony of *S. Sanguinis* GW2 was aseptically picked from a streaked blood agar plate and inoculated anaerobically in 100 ml of TSB for 72 hours on a shaking incubator at 37°C and 100 rpm. The cheek models were sterilised in 70% ethanol and thoroughly rinsed with sterile distilled water. Subsequently the models were aseptically fixed on the reactor rods of an anaerobical bioreactor with sterile threads and 30 ml of the inoculum was pipetted in the bioreactor which was then topped up with 0.5% sucrose-containing TSB. The bacteria were incubated in a batch phase for 24 hours at 37°C and 100 rpm before starting the continuous flow of fresh medium. After 5 days, 30 mL of an *S. Mutans* bacterial culture (prepared as described in section 2) was pipetted into the bioreactor, left for 24 hours in a batch phase and the mixed biofilm growth continued with fresh medium inflow for 3 more days. After the end of the experiment the models were aseptically taken out of the bioreactor, fixed, dehydrated and dried out with hexamethyldisilasane [11].

The two-species biofilm was imaged with a Nikon digital camera and more details were obtained with LM and SEM as revealed in Fig. 80. The cheek models were covered with dense white biofilm and the LM images showed white clusters of bacteria. SEM also showed dense bacterial population all over the vinyl strips.

5.   Summary

Models of the root canal, tooth and cheek were developed for the purpose of studying the cleaning of biofilms from surfaces such as teeth and dental implants, gingiva, cheeks, tongue, etc. *Serratia* species were utilized as a biofilm model because they easily grow on every material and their mineralization into biomimetic HA is possible making it potentially an excellent model of dental plaque and calculus (grown on the root canal model, flat glass and Ti). The oral bacteria *S. mutans* were used as representative of the most prevalent species in human plaque and the *in vitro* plaque was grown on flat glass and Ti surfaces, as well as on tooth models and gingiva. And finally, a two-species biofilm, consisting of *S. Sanguinis* and *S. Mutans* as a model of the microbiota of mucosal surfaces was grown on cheek models.

References

[1]   L. Hall-Stoodle, J. Costerton, P. Stoodley Bacterial biofilms: from the natural environment to infectious diseases, Nature Rev. Microbiol. 2 (2004) 95–108. https://doi.org/10.1038/nrmicro821

[2]   H. Medina Ledo, A. Thackray, I. Jones, P. Marquis, L. Macaskie, R. Sammons, Microstructure and composition of biosynthetically synthesised hydroxyapatite, J. Mater. Sci. Mater. Med. 19 (2008) 3419-3427. https://doi.org/10.1007/s10856-008-3485-3

[3]   P. Pattanapipitpaisal, A. Mabbett, J. Finlay, A. Beswisk, et al. Environ. Technol. 23 (2002) 731-745. https://doi.org/10.1080/09593332308618367

[4]   https://en.wikipedia.org/wiki/Calculus_(dental), last accessed December 2016

[5]   G. Tannok, Normal microflora. An introduction to microbes inhabiting the human body, Chapman and Hall, London, 1995, p.16

[6]   E. Theilade, Factors controlling the microflora of the healthy mouth, in: M. Hill, P. Marsh (Eds.), Human microbial ecology, Boca Raton, Fla, CRC Press, Inc., 1990, pp. 2–56.

[7]   S. Cohen, K. M. Hargreaves, Pathways of the pulp, 9th edition, Mosby Elsevier, 2006.

[8]   Q. D. De Deu, B. Horizonte, Frequency, location, and direction of the lateral, secondary, and accessory canals, J. Endodon. 1 (1975) 361–366. https://doi.org/10.1016/S0099-2399(75)80211-1

[9]   D. H. Pashley, S. M. Thompson, F. P. Stewart, Dentin permeability: Effects of temperature on hydraulic conductance, J. Dent. Res. 62 (1983) 956–959. https://doi.org/10.1177/00220345830620090801

[10]   V. Allan, M. Callow, L. Macaskie, M. Paterson-Beedle, Effect of nutrient limitation on biofilm formation and phosphatase activity of a Citrobacter sp., Microbiol. 148 (2002) 277-288. https://doi.org/10.1099/00221287-148-1-277

[11]   F. Braet, R. de Zanger, E. Wisse, Drying cells for SEM, AFM and TEM by hexamethyldisilazane: a study on hepathic endothelial cells J. Microsc. 186 (1997) 84-87. https://doi.org/10.1046/j.1365-2818.1997.1940755.x

[12]   AC. Thackray, Bacterial biosynthesis of a bone substitute material, PhD Thesis, University of Birmingham, 2005.

[13]   G. Tannok, Normal microflora. An introduction to microbes inhabiting the human body, Chapman and Hall, London, 1995, p.16.

[14]   B. Liu, K-L. Li, K-L. Kang, W-K. Huang, J-D. Liao, J. Phys. D: Appl. Phys. 46 (2013) 275401. https://doi.org/10.1088/0022-3727/46/27/275401

[15]   S. Suzuki, S. Imai, H. Kourai, Background and evidence leading to the establishment of the JIS standard for antimicrobial products, Biocontrol. Sci. 11 (2006) 135–145. https://doi.org/10.4265/bio.11.135

CHAPTER 6

Disruption of *In vitro* Dental Plaque Using the Energy of Cavitation

Abstract

Dental plaque and calculus on teeth can be removed by the dentist using hand instruments (periodontal curettes) or dental scalers [1]. The latters are a preferred form of treatment modality by dental health professionals. Periodontal scalers are hand-held dental instruments used in the prophylactic and periodontal care of human teeth, including scaling and root planning. The metal working end comes in a variety of shapes and sizes, but they are always narrow at the tip to access narrow spaces between teeth, as well as to clean periodontal pockets (Fig. 81). They differ from periodontal currettes, which possess a blunt tip. Scalers are not 100% efficient and in order to improve them it is essential to study biofilm growth, removal and structure. Ultrasonic scalers have been shown to be more effective than hand instrumentation for improving periodontal health. The debridement of plaque and calculus is primarily achieved by the mechanical chipping action of the scaler tip [2-4]. A stream of water flows over the tip to prevent frictional heating and to wash debris from the treatment site. Clinicians have identified generation of cavitation bubbles in the irrigation water around the oscillating ultrasonic scaler tip [5,6]. Cavitation involves the growth and collapse of microscopic bubbles (cavities) within a liquid, which can explosively collapse when critical conditions of pressure and temperature are reached [7,8]. Near a solid hard wall, bubbles tend to collapse in the direction of the wall. Alternatively, during bubble collapse next to a soft wall (like a biofilm covering a wall), the soft material might be pulled from the wall toward the bubble [8]. The collapse can lead to the breakdown of water and production of reactive species (radicals) inside the bubble [9]. In addition, collapse near a surface results in high-velocity microjets or microstreaming (hundreds of meter per second) that impact on a surface and aid surface cleaning [9-11]. The bubble collapse can also trigger a couple of consecutive bubble growths and collapses until it is damped out. The violent bubble collapse is called transient cavitation and is associated with surface erosion [12,13], surface cleaning [14] and other mechanical effects of ultrasonic cleaning. Recently, a new idea has risen in ultrasonic scaler debridement, which involves the use of cavitation bubbles as another mechanism to remove plaque and calculus from the teeth surface, periodontal pockets and dental implants and may lead to advances in enhancing the cleaning process without contact with the teeth [15-17].

Keywords

Plaque Disruption, Cavitation, Ultrasonic Dental Scaler, Scanning Laser Vibrometry, Sonochemiluminescence

Contents

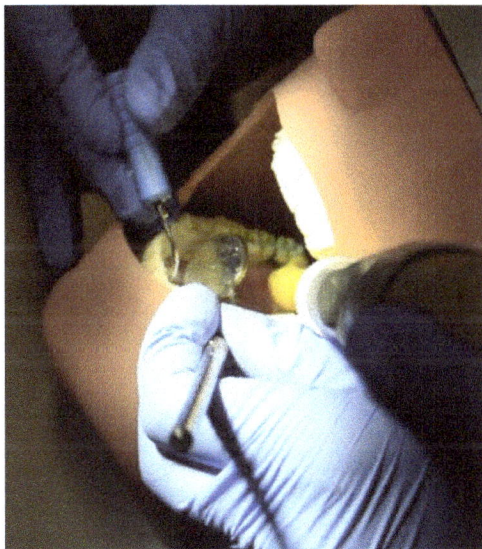

Figure 81. A dentist removing artificial plaque from a mannequin using a dental scaler.

1. Scanning Laser Vibrometry Studies of Ultrasonic Dental Scalers

Tip vibrational motion is not precisely understood and a better appreciation of the movement of ultrasonic scalers would enable clinicians to understand and improve their techniques during routine cleaning of the root surface. Understanding how ultrasonic scaling instruments oscillate when in use is essential to gain more knowledge regarding the cleaning mechanisms of the system. In the last decade, the oscillations of ultrasonic tips have been successfully assessed using an efficient and non-invasive technique, namely scanning laser vibrometry (SLV). SLV has made evaluating the oscillation characteristics of ultrasonic scaler probes a rapid process, providing a more detailed understanding of the way in which factors such as constraint, load and wear affect scaler vibrations and hence its performance [18,19].

The principle of the SLV is shown in Fig. 82. A low power eye-safe (class 2, < 1 mW) He–Ne laser beam is split into a reference beam and an object beam within the scanner head. The reference beam is internally reflected; the object beam is focused onto the oscillating target (i.e. the scaling tip) and then it is reflected back into the scanner head. The frequency of the reflected light is Doppler shifted, the magnitude of the shift being proportional to the instantaneous velocity of the object in the direction of the laser beam.

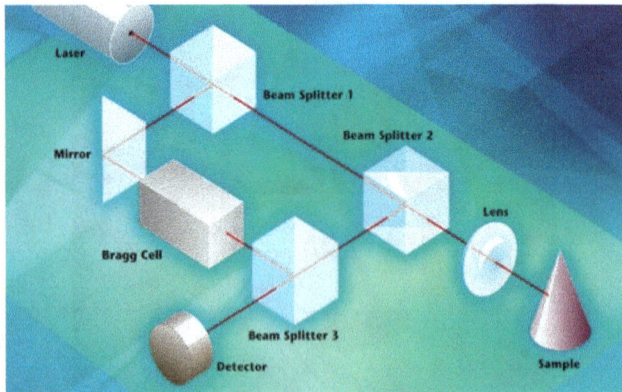

Figure 82. The principle of the SLV: the incident laser beam is split into a reference and an object beam, which then interfere.

The reference beam and the Doppler shifted object beam interfere within the scanner head. A photodetector records the interfered light and provides a voltage proportional to the velocity of the vibration. From this information the displacement amplitude of the vibration may be calculated (performed by the SLV software). SLV software enables the fundamental vibration frequency and subsequent harmonics to be readily observed. Vibration amplitude, frequency and spatial location are stored in the data manipulation system and are used to give full vibration mapping and animation capability.

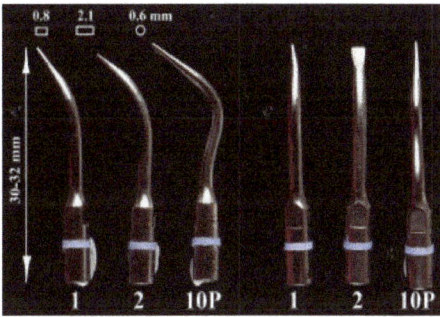

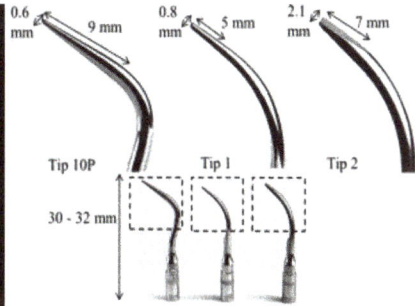

Figure 83. The front (left) and lateral (right) sides of the three scaler tips studied by SLV; tips cross-sections are also given.

Figure 84. Dimensions of the three scaler tips.

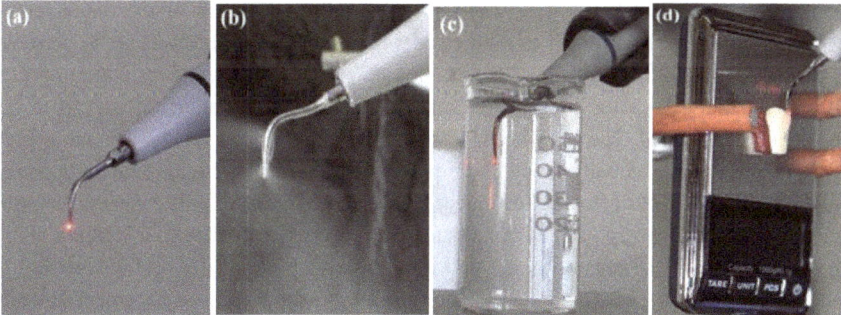

Figure 85. Tip vibrations are measured in air (a), under water flow (b), in water tank (c) and loaded (d).

This section aims to assess the vibration motion of a piezoelectric ultrasonic scaler with different shaped tips ranging from broad based shape to those which resemble a thin probe shape for better accessing periodontal pockets. These were assessed under simulated clinical conditions to determine which factors influence the movement the most.

A piezoelectric dental scaler (Satelec P5XS Newtron, Acteon, France, 30 kHz) was selected for this study with three tips of different designs (Figs. 83, 84) used by dental specialists to access subgingival pockets. Tips 1 and 2 had rectangular cross-section and tip 2 was broader than tip 1. Tip 10P had circular cross-section and was the thinnest, and the longest tip. Tips were handmade of stainless steel and before the analysis and experiments they were cleaned with ethanol. The analysis of tip vibrations was performed under unconstrained and unloaded conditions in air (Fig. 85 a). Subsequently, the oscillating tips were put under constrained conditions by adding water either as a flow from a small outlet on the back of the tip, running over its body at a constant rate of 30 ml/min (Fig. 85 b), or in a water tank containing 50 ml of distilled water (Fig. 85 c). Finally, the tips were tested against a plastic tooth attached to a digital balance (Timetop, capacity 1000 g, precision 0.1 g) and loads from 10 g to 150 g were applied towards the front side of the tip during its vibration (Fig. 85 d); contact was made at the free end of the oscillating tip (final 1 mm).

The vibration analyses of the scaler tips were performed using a SLV high frequency 2D scanning vibrometer system (Fig. 86; PSV 300-F/S, Polytech GmbH, Waldbronn, Germany), which works with an eye-safe He–Ne laser (λ=632.8 nm). A frequency measurement range of 0-100 kHz was selected to allow detection of the fundamental frequency of the scaler, as well as higher order harmonics. The SLV was set to perform fast Fourier transformation using 800 data points, giving a frequency resolution of 125 Hz thus enabling both the longitudinal and lateral oscillation pattern of the tip to be accurately determined. The laser beam was focused at the tip free end as shown in Fig. 85a and a number of equally-spaced scan points were chosen along the tip length, from the free end to as close to the handpiece as could be measured. The handpiece was clamped so that the tip was vertical and clearly visible to the SLV camera. Full characterization of the unconstrained tip vibrations was performed with the laser beam focused on the front, back and side of the tip. The maximum tip displacement at each scan point was measured and an average of 5 measurements was recorded for each tip, at each power setting (1 = lowest to 20 = highest). Each scan lasted approximately 10 sec with an interval of 20 sec between scans.

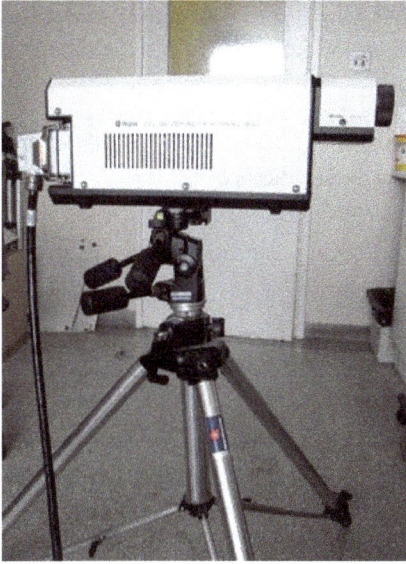

Figure 86. The SLV device used for measurement of the scaler vibrations.

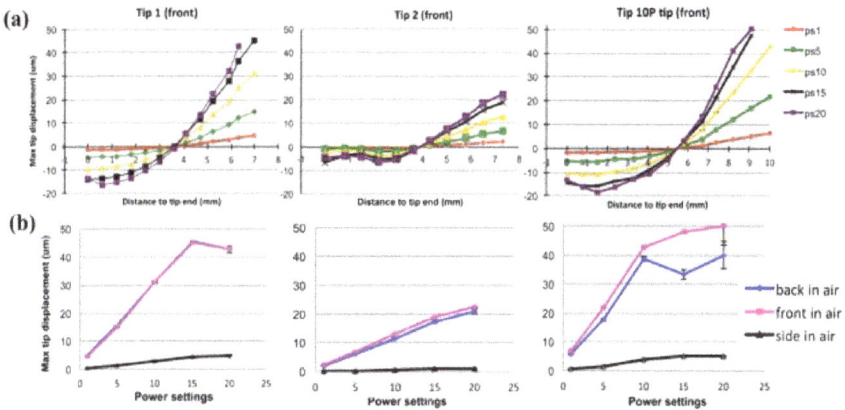

Figure 87. (a) Maximum displacement amplitude (± SD, μm) along the length of the tip vibrating in air, plotted as a function of the distance from the tip unconstrained end; (b) comparison of the displacement amplitude (± SD, μm) measured for the front, back and side of the tip free end vibrating in air, plotted as a function of power settings.

163

SLV data from the scan point located at the unconstrained end of the tips were analysed using IBM SPSS v.21 for Windows (SPSS Inc., Chicago, IL, USA). The significance of variation in the maximum tip displacement amplitude at different generator power settings, as well as under various constrains and load conditions was tested using univariate analysis of variance (ANOVA, general linear model) and multiple post hoc comparisons (Tukey test). For similar studies the significance level is set as $p < 0.05$ with the dependent variable being the displacement amplitude.

The SLV data is presented as the maximum displacement amplitudes along the length of the unconstrained tip and the scans revealed one node at about 3.5 mm (tips 1 and 2) or 4.5 mm (tip 10P) measured from the tip free end (Fig. 87 a). Two antinodes could be identified: at the very end of the tips (first scan point) and at about 6.5 mm, 5.0 mm or 8.5 mm for tips 1, 2 and 10P, respectively. Displacement clearly increased with increasing power from the lowest (1) to the maximum generator power (20). There was an over-range signal when recording the tip end at the highest power settings. Tip 2 exhibited the lowest displacements whilst tip 1 gave higher amplitudes of oscillation. Tip 10P had the highest vibrations at all power settings. Side vibrations were also assessed (Fig. 87 b) and they consisted of displacements of about 10% of the values measured for the front of tips 1 and 10P and 5% for tip 2 at high power settings (15 and 20).

Tip oscillations under water flow were compared to those in air (Fig. 88) and it was found that the displacement amplitude decreased for both the tip front and back when water flow was present. At average power setting (10) the water flow decreased the tip 10P vibrations. Contrary to the effect of the irrigation, when water was used as a volume constraint, the vibrations of the three tips were unambiguously higher at settings 1 to 10 than in air for both measurements of the tips front and back, with tip 2 showing the most stable behaviour for all power settings.

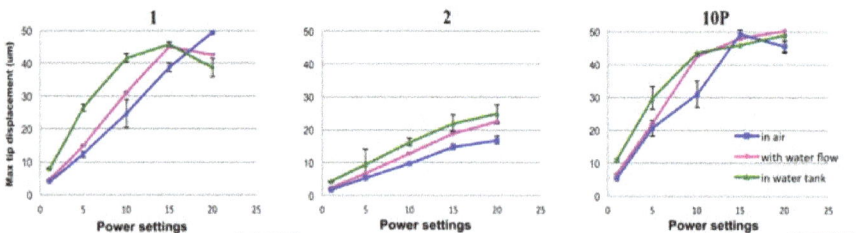

Figure 88. Comparison of the effect of water constraint on the displacement amplitude ($\pm SD$, μm) of the tip free end (front side).

The maximum displacement amplitude at the tip free end under various loads was plotted as a function of the generator power settings (Fig. 89). The amplitude steadily decreased for all three tips with increasing load from 10 g to 150 g. Tip 2 had the most stable behaviour upon loading and yielded the lowest displacement values. Difficulties in measuring the displacement at the highest power were experienced for tip 1 at 100 g and 150 g, and especially for the thinnest tip 10P at all loads. The increase of the generator power yielded a linear increase in the displacement amplitude with the maximum value measured for setting 15.

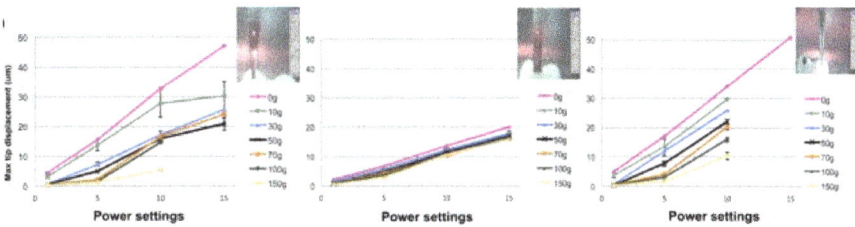

Figure 89. Comparison of the effect of loads on the displacement amplitude (± SD, μm) of the tip free end (front side).

When used for dental scaling, the ultrasonic scaler is set at an arbitrary value often being "tuned" by the operator to obtain the best sound before using the instrument on the patient. Previous work [16] has shown that there is variation with ultrasonic scalers and this may produce different clinical outcomes. The majority of studies do not perform a basic test of the operating scaler. The tip may undergo a range of oscillatory movements depending on the tip being in air or water. The latter also leads to differences depending if immersed in water or if water is passed over it.

By using laser vibrometry it was possible to carry out detailed analyses of the oscillations of piezoelectric dental tips under different conditions (unconstraint, constraint by water flow or water volume and loaded). We could also observe the tip response to varying generator power, as well as to determine the locations of nodes and antinodes along the length of the scaler tip which is associated with the occurrence of cavitation [15] and thus can be related to the disruptive effect of the tip when approaching biofilms.

Performing unconstrained vibration measurements of a ultrasonic tip (i.e. in air) gives a guide to the performance of the power generator driving the tip and also to the way in which the tip responds to changes in power. Previously the increase of power settings has led to the increase of the tip displacement values [20] and this was again confirmed in this study. Also, the vibrational antinode with the greatest displacement amplitude was found to always occur at the free end of the tip.

Tip design is influential in determining the tip vibrations [15,18]. Tips 1 and 2 had a flat cross-section and tip 2 was broader than tip 1. Tip 10P had a circular cross-section, it was thinner and longer than the other two tips (Fig. 83). The broadest tip 2 had the lowest displacement values and the most stable vibration motion while tips 1 and 10P are thinner and had higher amplitudes of oscillation (Figs. 87-89) especially the longest, 10P. Even though thinner tips are preferred in clinics because they enable easier access to tooth roots, the cavitation occurrence around thin tips may be reduced when used in areas with difficult access [15,21].

This study resembled the clinical situation where irrigation liquid or disinfectant water is flowing over the tip [15]. When the tip was vibrating under a water flow (Fig. 88), the oscillations were suppressed in comparison to those in air.

The movement characteristics of the tip in a large water volume have not been reported previously. However, liquid volumes are also often used when imaging dental tip vibrations and resulting cavitation bubbles with high-speed cameras [7,15]. Contrary to the measurements with the water flow, tip vibrations in the water tank increased and the displacement values measured for the tip front and back were generally higher than these in water flow or in air (Fig. 88). We consider that this boosting effect was due to the presence of a water volume around the tip, larger than the tip volume, which acted as an amplifier. The effect was clearly pronounced for tip 2 at all power settings, for tip 1 up to setting 15 and for the back of tip 10P at all power settings. Clinically there will be much variation as the tip is rarely immersed fully in liquid unless working in the posterior part of the mouth. In the anterior region, the water flow is likely to move off the tip during usage.

The application of in vitro load was also found to influence the vibration patterns [18]. All tip displacements at settings from 1 to 20 decreased steadily with the application of the lower loads (< 70 g), as well as when it was increased to clinical loads, i.e. > 100 g (approximately 1 N) (Fig. 89).

Analysis of vibration data from the unconstrained end of all tips showed that power settings, environmental conditions and position of measurement were statistically significant variables ($p = 0.000$). Low (1, 5) and average (10) power settings were different from high (15, 20) power settings ($p = 0.001$) while settings 15 and 20 were not different ($p = 1.000$). Displacements of tips 1 and 2 in air and water flow were different from those in the water tank ($p=0.0001$) and vibrations measured from the front and back of tips 1 and 2 were different from those from the side (p = 0.000). Post hoc comparison for tip 10P revealed differences for power settings 1 and 5 (p = 0.0001), while no differences were found for power setting 10 to 15 (p = 0.997), 10 to 20 (p = 0.717) and 15 to 20 (p = 0.886). Vibration data in air for tip 10P were significant from those under water flow (p = 0.006) and in the water tank (p = 0.0001), as well as the displacements measured from the front, back and side of this tip (p = 0.0001).

When load was applied to the tip, significant difference was found for the loaded 1, 2 and 10P tips in comparison to unloaded conditions (p = 0.0001). Tip vibrations due to loads > 70 g were also found different. A Tukey test for the three tips showed that tip and power settings were significant variables (p = 0.000).

The degree of variability observed in the unconstrained and constrained vibrations of tips 1 and 10P might be due to several factors. One reason is when the energy of the working frequency is transferred to a harmonic frequency, which results in a decrease in the displacement amplitude measured for the fundamental frequency. Analysis of the frequency spectra obtained with the SLV software up to the second harmonic frequency revealed that this was probably not the case in this study. Other reasons could be variability in the behaviour of the piezoelectric handpiece driving the tip, which may arise because of differences in tightness of fit when they were screwed into position in the handpiece, even though care was taken to screw the tip correctly. Wearing out of the tip could also has been a significant factor affecting the decrease of the oscillation amplitude. The SLV data obtained in this work gives us the opportunity to obtain better knowledge on tip motion in constrained and loaded environments. They will also allow dental specialists to better understand the cleaning mechanisms of ultrasonic scalers and thus to improve their techniques during routine cleaning of teeth. It may also lead to improvements in tip design and to the production of more effective dental instruments for clinical use.

2. Cavitation and its Disrupting Efficacy

Root canal treatment becomes necessary when the pulp within a tooth becomes infected (through bacteria or their by-products) [22,23]. It is undertaken using a combination of physical instrumentation and irrigation with chemical irrigants such as sodium hypochlorite [24]. Root canal anatomy is complex and complete disinfection of the root canal through instrumentation alone is not possible. A biofilm may remain on canal walls together with hard and soft tissue debris. Such material may be present in more inaccessible areas such as fins or isthmuses between canals. Irrigation is the key to reaching into and cleaning out such areas [24]. Ultrasonic instruments are known to produce biophysical forces, such as cavitation and acoustic streaming fields in the irrigant around the tip and these may assist in the disruption of bacteria and moving irrigant around the canal [25,26].

The second stage in this work was to assess the effect of the cavitation occurring around the vibrating ultrasonic tips on the disruption of model biofilms used to mimic the dental plaque formed on flat or confined surfaces (teeth front and back, interdental spaces, root canals, isthmuses between canals, etc.). It was shown in chapter 5 that *Serratia* sp. were able to form biofilm within the canals of a root canal model and the biofilm was successfully mineralized in an appropriate medium to resemble the calculus found on human teeth (Figs. 71-74). The biofilm and calculus removal through a new generation of dental instruments that use cavitation to disrupt the biofilm was targeted. Cavitation bubbles around a scaler tip have been expected to remove plaque and calculus from teeth surfaces due to the shock waves produced upon their collapse [5-8]. Can this cavitation disrupt biofilms successfully leading to removal of infection, therefore allowing the body to repair itself? Envisaged disruption of the biofilm by ultrasonic dental instruments will provide a tremendous clinical advantage in periodontal and endodontic treatment, leading to a new range of instruments, which can be extended to cleaning hardly accessible places.

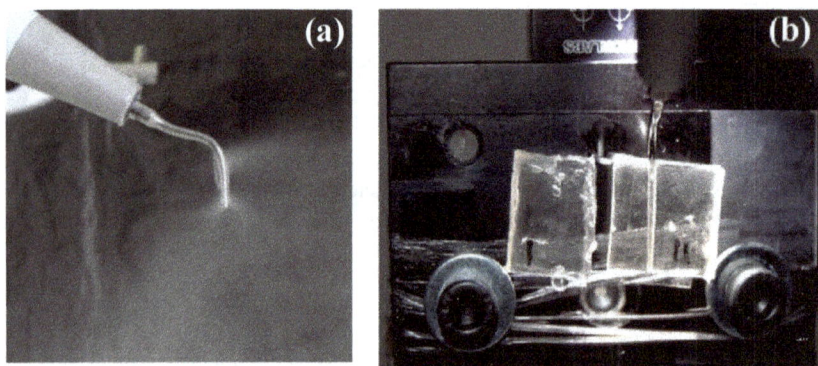

Figure 90. (a) Image of the ultrasonic handpiece with Satelec tip 10P working under a water flow and (b) the set-up used for biofilm disruption within the silicone root canal model.

Tip 10P (see Figs. 83 and 84) working with water flow over the tip was used to disrupt the *in vitro* plaque, as well as the mineralized biofilm inside the root canal model (Fig. 90). The silicone models were fixed in a vertical position on a 3D micropositioning stage (Thorlabs, USA) and the handpiece was attached to a stand. By moving the 3D stage with the models, the tip free end was inserted in the main canal as shown in Fig. 90 b. The tip oscillation was started and the interaction lasted for 30 or 60 sec by using fixed power setting of 15 with irrigation. SEM revealed that at these experimental settings the plaque and mineralized biofilm remained on the canal walls and no visible differences were found before and after the contact with the tip (Fig. 91). Most probably due to the complexity of the canals and their small dimensions the thick scaler tip ($\varnothing$ 0.6 mm) was not successful in removing the deposits without using the mechanical contact with the canal walls. We have also not imaged the process of cavitation generation within the canals of the silicone model at this stage, which is possible only with a high speed camera. Whether or not cavitation will occur depends not only on the acoustic field but also on physical properties and conditions such as the amount of gas in the liquid and the presence of nucleation sites [27]. We do not know at this stage if the conditions for occurrence of cavitation have been fulfilled. Similar work of Macedo *et al.* [28] demonstrated the use of a hydrogel with viscoelastic behavior and mechanical properties comparable to real biofilms, which was used to cover the canals of the silicone model.

Thinner files ($\varnothing$ 0.25 mm) for accessing root canals were used with ultrasonic water or NaOCl irrigation to remove the hydrogel. These initial results showed that the hydrogel could be pushed off the canals of the model and also the ultrasonic irrigation improved the hydrogel removal from the canal anatomy, but the creation of stable bubbles on the hydrogel-liquid interface may have reduced the cleaning rate. A greater depth of cleaning was achieved from an isthmus than from a lateral canal and at a faster rate for the first 20 sec. The effect of the ultrasonic irrigation was reduced when stable bubbles were formed and trapped in the lateral canal. Different removal characteristics were observed in the isthmus and the lateral canal, with initial highly unstable behavior followed by slower viscous removal inside the isthmus. This work shows that in our experiment the tip might have been too thick and the interaction time too short to cause any visible removal of the biofilm. On the other side, Macedo *et al.* have not investigated the hydrogel adhesion to the silicone model in order to compare it with the strongly adhesive bacterial films. When it comes to adhesion, the properties of the substrate, such as for example surface roughness and hydrophobicity are also important [27,29]. The hydrogel in their work was not grown but deposited onto a glass slide and therefore its attachment to the substrate is expected to be generally weaker than for a biofilm, relying only on physicochemical interactions between the glass and hydrogel proteins [30], which also aid its removal with the scaler files.

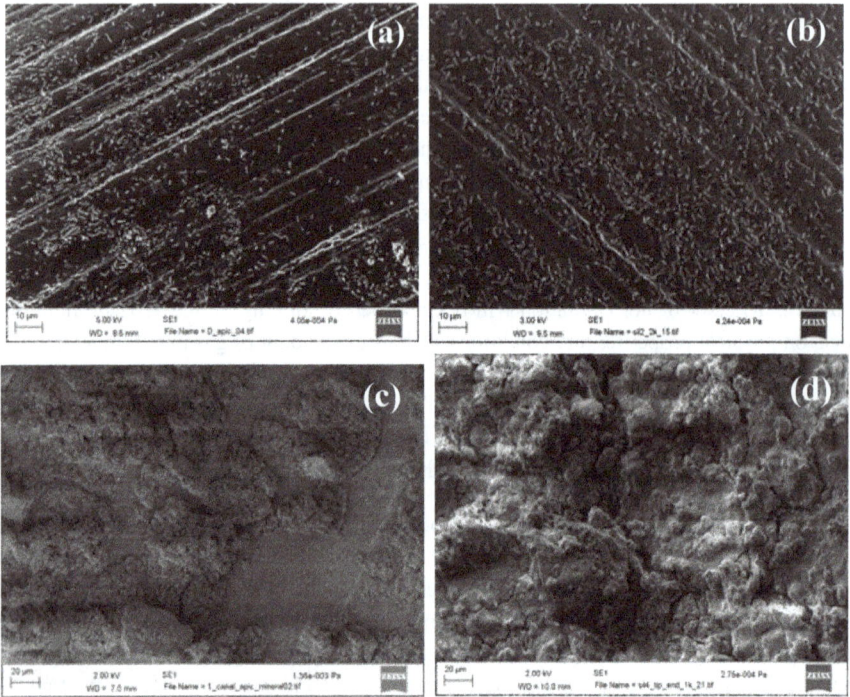

Figure 91. SEM images of (a, b) the silicone models with biofilm in the main canal, before and after contact with the tip (scale bar 10 μm); (c, d) the silicone models with mineralized biofilm before and after contact with the tip (scale bar 20 μm).

Figure 92. (a) Two positions along the tip length were used to approach the biofilm with (A = tip free end, B = tip bend); the set-up used for biofilm disruption on glass (b) and Ti (c).

In the experiment presented further, *Serratia in vitro* plaque was disrupted on flat glass and Ti surfaces using the three tips whose oscillations were fully characterized with the SLV as described in section 1 of this chapter. Ultrasonic scaler debridement, which can involve the use of the cavitation bubbles, may enhance the cleaning process of plaque and dental calculus in periodontal pockets and around dental implants as the high energy levels generated by the ultrasonic scaler will disrupt or disaggregate bacteria within the plaque [15-17]. In particular, periodontal bacteria have been shown to be sensitive to ultrasound [31,32]. In a deep periodontal pocket not all the bacteria will be exposed to the same magnitude of shear forces, thus possibly producing a further clinical variability. Previous work with a model system composed of particulate matter removed from a glass slide showed that activity in the associated cooling water cleaned the slide close to the probe [33]. This removal was demonstrated on the plaque-covered surface of the teeth but no direct measurement of the amount removed was made [34]. Similar areas of erosion or "cleaning" were observed on root surfaces subjected *in vitro* to a scaling tip with water flow [35]. Most of the studies in the literature utilizing biofilm removal with scalers report direct mechanical contact with a biofilm-coated material (teeth, HA or Ti disks) [17,36-39]. However mechanical contact between the scaler tip may yield hard tissue defects [36,37,39]. Avoiding the mechanical contact with the tip will bring comfort to the patient due to the lack of vibration, pain and noise experienced during scaling.

Tips of differing shape and width (standard Satelec tips 1, 2 and 10P, shown in Figs. 83 and 84) with water irrigation from a small outlet on their back were used to disrupt the biofilm. Tips were screwed to the handpiece of a Satelec scaler that was fixed on a XYZ micropositioning optical stage, which allowed precise setting of the tip-biofilm distance of 40 μm or 1 mm. The scaler was operated at a fixed power setting giving a tip longitudinal displacement of 15 μm for each tip. Fixed and dried biofilm-coated glass or Ti disks were attached to a thick glass plate (130x50x5 mm), which was clamped on a stand. Direct mechanical contact of the tips with the substrates was avoided by placing them at the minimal possible distance of 40 μm from the tip (Fig. 92). The tips were placed with their front side parallel to the biofilm and their free end was facing down. The scaler was run for 30 sec with water irrigation (30 ml/min). Two positions along the tip length, associated with the main areas of cavitational activity were consecutively used to approach the biofilm: point A (the tip free end; Fig. 92 a) and point B (the tip bend). After scaling, remaining biofilm debris was removed from the disks by gentle manual rinsing with distilled water. The models were left to dry in air at room temperature. To extract the effect of water impingement on the biofilm disruption, a stream of water from a 50 ml syringe was directed with a speed of 15 or 60 ml/min to the biofilm placed at 1 or 5 mm distance as a control.

The biofilms before and after scaling with the ultrasonic tips were sputter-coated with gold for 2 min then imaged with SEM (Zeiss EVO MA-10, Cambridge, UK, 5 kV, 10 mm working distance). LM was also carried out (Nicon Eclipse TE300, Japan and GE-5 digital microscope, View Solutions Inc., USA). A digital camera Nikon D40 was used to obtain overall images of the biofilms and quantitative estimation of the disrupted areas was performed using ImageJ image analysis software [40-42]. Disrupted areas from the five repetitions at each setting were estimated in pixels using a threshold method on an 8-bit smoothed image in ImageJ, which showed the biofilm as white pixels and the disrupted area as black pixels [41]. Statistical differences of the removed plaque were analysed using 1-way univariate analysis of variance (ANOVA) and multiple post hoc comparisons (Tukey test) at a significance level of $p < 0.05$, with the dependent variable being the disrupted biofilm area. IBM SPSS statistical software (SPSS Inc., Chicago, IL, USA, v.21) was used.

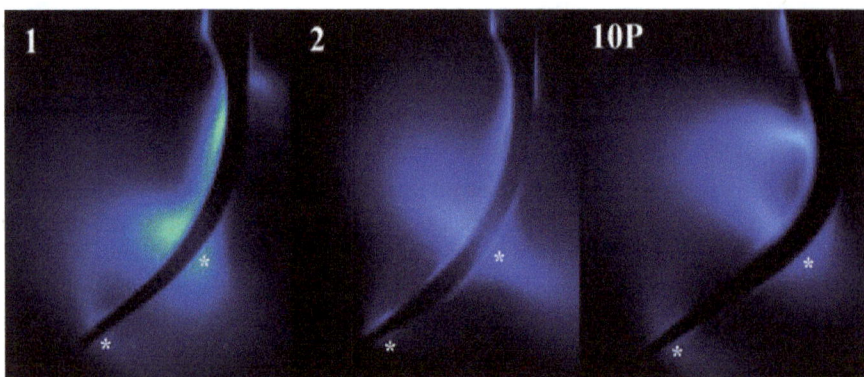

*Figure 93. Two main areas of cavitational activity along the tips (marked with * and corresponding to positions A and B in Fig. 92 a) were revealed by monitoring the SCL emission from luminol solution and used in approaching the biofilm during scaling.*

The cavitation activity of the ultrasonic tips was assessed by monitoring the emission of sonochemiluminescence (SCL) from luminol solution. Luminol solution (5-amino-2,3-dihydro-1,4-phthalazinedione) was prepared in deionised water at a concentration of 1×10^{-3} mol/l luminol, 1.10^{-4} mol/l hydrogen peroxide and 1×10^{-4} mol/l ethylenediamine tetraacetic acid and adjusted to pH 12 by addition of sodium hydroxide [43]. A Canon EOS 500D digital single-lens-reflex camera with a Canon EF-S 60 mm f/2.8 USM Macro lens set at ISO 1600 was used to record the emission. The apparatus and camera were placed in a light-proof box. 40 ml luminol solution was poured into a quartz glass fronted cuvette (25x25x75 mm) and the tip was fixed inside the cuvette in position side-on to the camera. The scaler was operated for 15 seconds with long exposure photograph. Each tip was photographed five times at power setting of 15. The intensity of the luminol emission was calculated and further image post-processing was performed using image analysis with ImageJ [40]. A background image was taken with the ultrasound switched off and subtracted from the SCL image.

Cavitation bubbles around scaler tips might be expected to remove plaque and calculus from teeth surfaces due to the shock waves produced upon their collapse [7,9]. In order to assess the role that cavitation plays in deposit removal from teeth, it is first necessary to identify the distribution of cavitation around each tip, so that possible differences due to variations in tip design can be elucidated. One method to investigate the spatial distribution and the degree of cavitation occurring around ultrasonic scaler tips is to monitor the SCL emission [30]. Studies using SCL have proven that when cavitation occurs around ultrasonic scalers, it happens in clusters and not along the whole length of the tip [44-46]. The cavitation fields (Fig. 93) produced by each of the tips 1, 2 or 10P in the luminol solution are shown in Fig. 93. SCL arising from the reaction of luminol with ultrasonically generated hydroxyl radicals was detected as intense regions of activity surrounding the bend of each tip while there was little activity at the tip free end. As expected, higher power settings produced the greatest intensity, and therefore the greatest amount of cavitation for each tip [15]. Therefore only the images obtained at power 15 of the scaler are shown here.

In this study, SCL showed little cavitation at the tip free end and the low intensity of the luminescence made it difficult to analyse the extent of cavitation there (Fig. 93). On the other side, a strong antinode has been detected at the tip end with SLV [47]. SLV shows the antinodes corresponding to the points of largest tip vibration (equivalent to highest displacement amplitude) and they are associated with the cavitation generated around the tip. SLV has been successfully used before to analyse the vibrational motion of the scaler tips under study in this work [47]. It revealed that the highest vibration occurred at the tip free end and there was another vibration point close to the tip bend, which correlated well with the position but not with the SCL intensity of the cavitation bubble cloud seen in Fig. 3. High-speed imaging (HSI) has also been used to characterize cavitation in fields such as mechanical and biomedical engineering [48-50], and in dentistry [45,46,51]. It can directly image the bubble occurrence instead of using the indirect SCL and it displays activity in real time with higher resolution, allowing image analysis to be done on individual bubbles and clusters. Using HSI we confirmed the occurrence of cavitation both around the tip bend and tip free end, and the latter was substantial for the three tested tips [52] contrary to the SCL. Our work also confirmed that cavitation occurs at the antinodes, or points of highest displacement of ultrasonic scaler tips, with no or small cavitation occurring at the nodes [52,53]. This was attributed to the higher amount of liquid being displaced by the scaler at the points where the amplitude of vibration is largest [53]. According to the linear wave theory an antinode (or node) for displacement is always an antinode (or node) for pressure [54]. Therefore, the lowest pressure occurs at an antinode for displacement, which causes the cavitation there.

Combining data from the three different techniques allowed us to understand the link between scaler vibrations and the occurrence of cavitation. We investigated the cavitation activity around the tips with luminol photography, HSI or SLV in order to determine the spatial distribution of cavitation and chose the most effective for biofilm disruption points along the tip length (points A and B in Fig. 92 a). LM and SEM images of the *Serratia* biofilms grown on flat glass surfaces were shown in Figs. 72 a and c. Typical images of the disrupted biofilms after they have been approached with the three tips at 40 μm or 1 mm distance are shown in Fig. 94. The disrupted areas had circular or oval shapes when the tip end was used and an irregular shapes when the tip bend approached the samples.

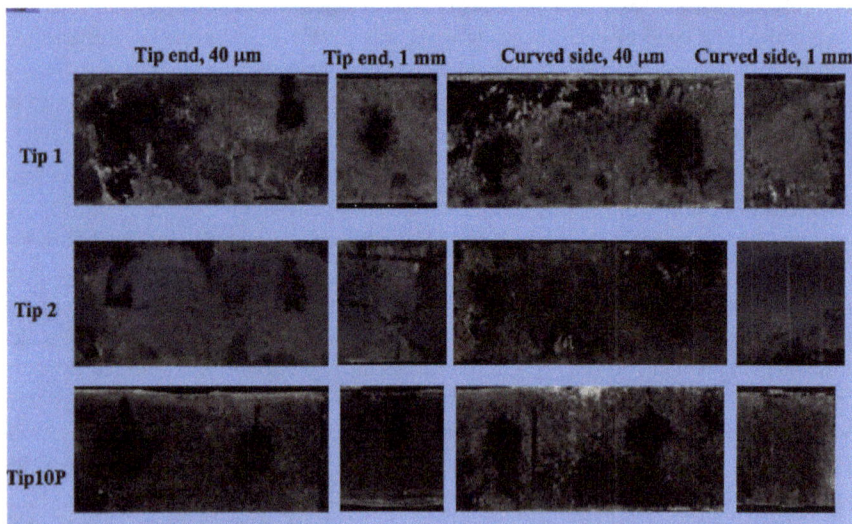

Figure 94. Serratia biofilm removed by the three tips 1, 2 and 10P from flat glass; images obtained with a Nikon digital camera.

SEM analysis revealed that the *Serratia*-coated glass substrates were covered with numerous bacteria (Fig. 95 left). After disruption with the scaler, the biofilm was mostly removed with fewer bacteria left on the surface (Fig. 95 right).

Table 10. Disrupted Serratia biofilm area ($\pm$SD, mm^2) on glass slides

Satelec tip No	Tip free end (position A)		Tip bend (position B)	
	40 μm	1 mm	40 μm	1 mm
1	4.6 ± 2.2	3.4 ± 0.1	4.4 ± 1.0	-
2	4.0 ± 0.9	1.6 ± 0.2	6.1 ± 3.1	-
10P	3.3 ± 1.5	1.4 ± 0.5	2.8 ± 0.1	-

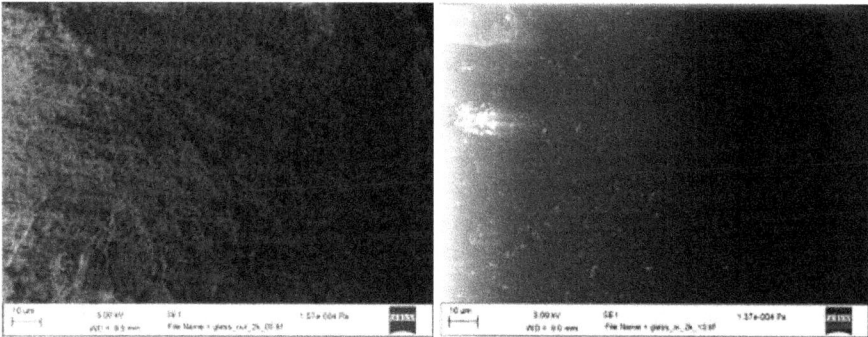

Figure 95. Typical SEM images of the Serratia biofilm on glass before (left) and after (right) the removal by scaling.

No unambiguous conclusion about the efficacy of the scaling can be made by only observing the images so the biofilm disruption was quantified using ImageJ software as described above and the results are presented in Table 10. None of the tips induced a lesion at all when placed at a distance of 1 mm. The most effective disruption amongst the tree tips was obtained with the bend of the wide tip 2 placed at the distance of 40 μm. Tip 1 was almost equally effective when approached the biofilm with both its free end at 40 μm and 1 mm distance, as well as with its bend at 40 μm. The closer the tips 2 and 10P were placed to the biofilm (i.e. at 40 μm), the better results were obtained with their free end or bend. When the tips were oriented side on, no lesion was obtained and only their "contact footprint" was visible. There was no disruption or footprint with the tip placed perpendicularly to the glass substrate, with only the end in contact with the biofilm (images not shown). The control experiment with water impingement from a syringe placed at a tip-biofilm distance of 1 or 5 mm also did not remove any biofilm.

Ti disks were also used as substrates for the biofilm growth due to the application of Ti as dental implants. Both biofilms were grown in the bioreactor. Reproducible bacterial biofilms (*Serratia* and *S. mutans*) with a thickness of about 2 µm were grown on Ti disks (grade 2, diameter 15 mm) using the procedures described in chapter 5. Five disks with *Serratia* or *S. mutans* biofilm were prepared for each setting (tip shape, position A or B, tip-biofilm distance 40 µm or 1 mm). Imaging, quantification and statistical analysis were performed in the same way as for the *Serratia*-covered glass.

Representative images showing the intact biofilm and disrupted areas on the Ti surface after approaching the *Serratia* and *S. mutans* biofilms with the tip front side are shown in Figs. 96 and 97. The biofilm was imaged with a digital camera (Figs. 96 a, 97 a) and more details were obtained with the LM and SEM. The disrupted areas had circular or oval shapes when the tip end was used and irregular shapes when the tip bend approached the samples. LM images obtained before the scaling showed that *Serratia* or *S. mutans* biofilms completely covered the surface of the Ti disks (Figs. 96 b, 97 b). SEM analysis revealed that the disks were covered with numerous bacteria (Figs. 96 c, 97 c). After disruption with the scaler, SEM and LM images showed that the biofilm was mostly removed with only a few bacteria left on the surface. The quantitative results of biofilm disruption are shown in Fig. 98. Both bacterial biofilms were most effectively de-scaled with the bend of the thinner tips (1 and 10P); the wider tip 2 did not induce a lesion at all when placed at a distance of 1 mm. Neither the tip free end, nor the bend of tip 10P cleaned *Serratia* biofilm at a distance of 1 mm. On the other hand, this tip showed the highest disrupting efficacy of *S. mutans* biofilm compared with tips 1 and 2 for all settings (positions A and B; tip-biofilm distance). The water impingement from the syringe placed at a tip-biofilm distance of 1 or 5 mm also did not yield any disruption of the biofilms grown on Ti.

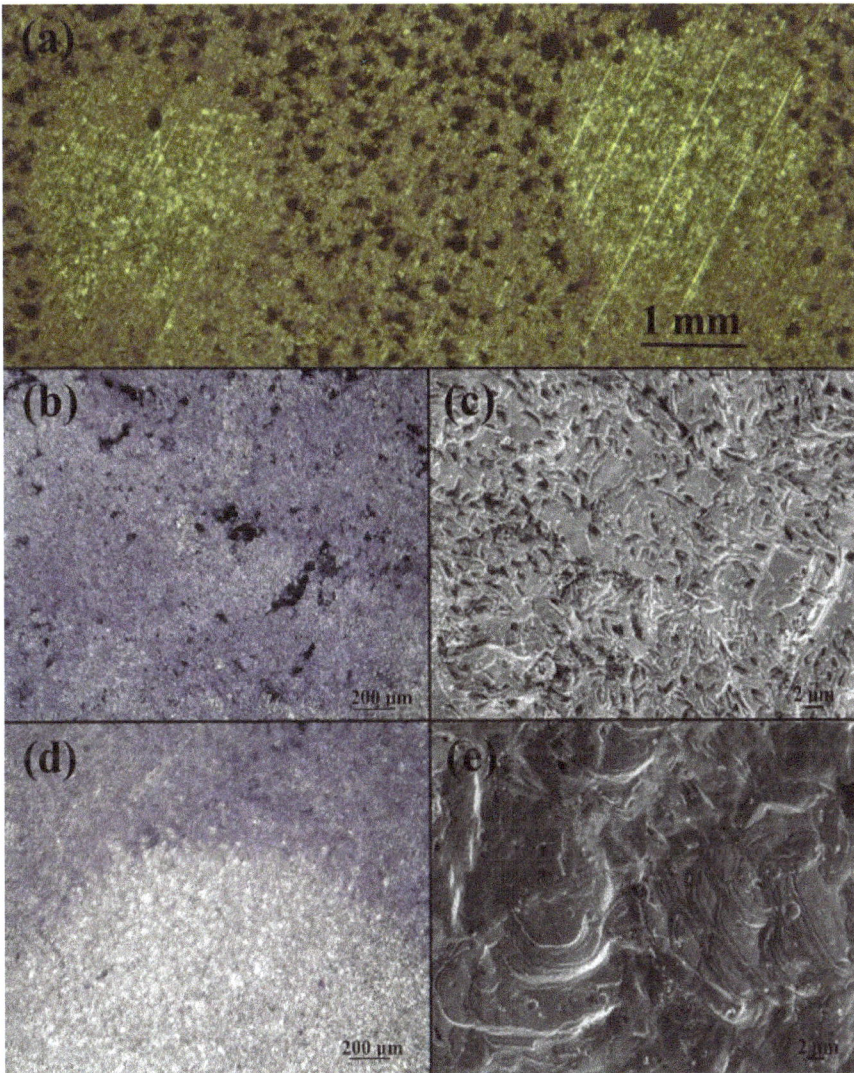

Figure 96. (a) Digital camera image of two disrupted areas of Serratia biofilm on a Ti disk using the free end of tip 10P at 40 μm distance; (b) and (c) LM and SEM images of the intact biofilm; (d) and (e) LM and SEM images of the disrupted biofilm (the lighter area in (d) is the disrupted biofilm).

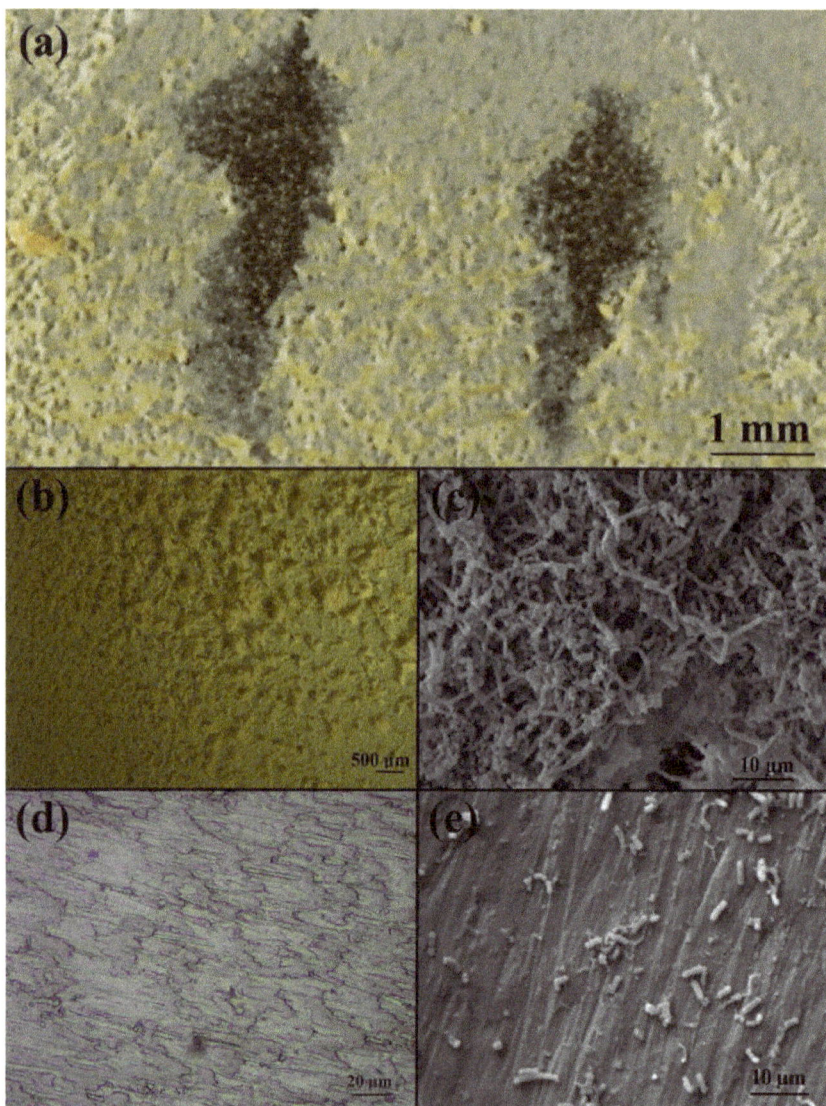

Figure 97. (a) Digital camera image of two disrupted areas of S. mutans biofilm on a Ti disk using the bend of tip 1 at 40 µm distance; (b) and (c) LM and SEM images of the intact biofilm; (d) and (e) LM and SEM images of the disrupted biofilm.

Serratia and *S. mutans* biofilms provided useful models for investigating the efficacy of the ultrasonic dental scaler for biofilm removal. The three tips removed the biofilms only when their front side was used due to the fact that the tip displacement and hence cavitation were stronger in the longitudinal direction of the tip vibration. There was insignificant tip displacement in the lateral direction as shown with the SLV study so the tip sides or tip end did not yield a lesion under the conditions used here. According to the data in Fig. 98, at 40 µm distance to the biofilm tips 1 and 10P were disrupting both biofilm models significantly better with their bend than with the free end ($p < 0.05$). The bend of the wider tip 2 was scaling off the biofilms as well as its end. Not all tips placed at the higher distance (1 mm) were successful in disrupting the biofilm models. At 1 mm, the bend of tip 1 also yielded significantly higher disrupted *Serratia* biofilm area compared to its end ($p < 0.05$); there was no disruption of both biofilms with tip 2 and no statistical difference was found for the efficacy of the bend of 10P over its end in scaling off the *S. mutans* biofilm ($p > 0.05$).

(a) *Serratia* biofilm **(b) *S. mutans* biofilm**

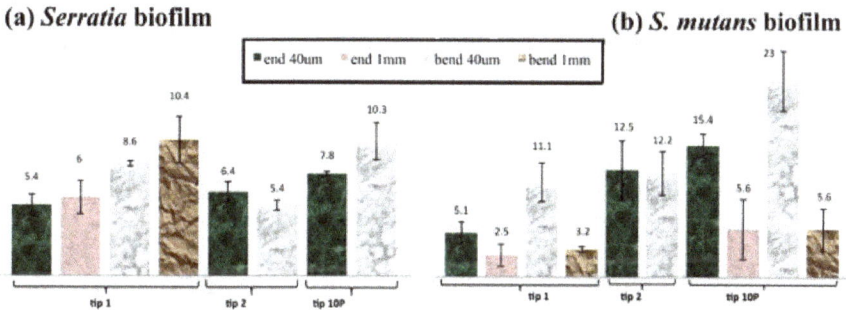

Figure 98. Serratia (a) and S. mutans (b) biofilm area disrupted with the three tips (mm^2).

3.   Occurance of Cavitation in Dentistry – Summary

Generation of cavitation is complex and displacement amplitudes may not be the only factor in the generation of inertial cavitation. The control experiment with water ejected from a syringe (no bubbles) showed no biofilm lesions and thus excluded the possibility that the water flowing over the tip itself disrupted the biofilm. Since the vibrating scaler tips were placed at a minimal distance of 40 µm to the biofilm and thus the mechanical chipping action was avoided, it was assumed that the cavitation bubbles cloud observed around the tips played a role in the *Serratia* and *S. mutans* biofilm disruption on glass and Ti flat surfaces. The contribution of streaming can be ruled out based on the above reported observation that there is no cleaning by streaming only (without bubbles). It should be noted that these tests were carried out with dried-on biofilms, which are likely to be more tenacious than fresh ones. Therefore the ability of the tips to remove biofilm in non-contacting mode under these conditions is even more impressive. Tip design, area of cavitational activity and tip-biofilm distance all influenced the biofilm disruption with the ultrasonic scaler. The effect of the cavitation bubbles around the scaler tips on the biofilm removal may lead to advances in enhancing the cleaning of teeth and dental implants with ultrasound scalers, avoiding mechanical contact with teeth. It is difficult to maintain the tip-tooth distance to less than 1 mm when holding the handpiece by hand and it will be a challenge for the clinicians, however it will bring benefits to the patients.

References

[1]   J. Nield-Gehrig, Fundamentals of periodontal instrumentation & advanced root instrumentation, Wolters Kluwer/Lippincott Williams & Wilkins, 6th edition, Philadelphia, PA, 2008, pp. 288-309

[2]   A. Badersten, R. Nilveus, J. Egelberg, Effect of non-surgical periodontal therapy. Moderately advanced periodontitis, J. Clin. Periodont. 8 (1981) 57–72. https://doi.org/10.1111/j.1600-051X.1981.tb02024.x

[3]   D. Breininger, T. O'Leary, R. Blumenshine, Comparative effectiveness of ultrasonic and hand scaling for the removal of subgingival plaque and calculus, J. Periodont. 58 (1987) 9-18. https://doi.org/10.1902/jop.1987.58.1.9

[4]   B. Loos, R. Kiger, J. Egelberg, An evaluation of basic periodontal therapy using sonic and ultrasonic scalers, J. Clin. Periodont. 14 (1987) 29–33 https://doi.org/10.1111/j.1600-051X.1987.tb01509.x

[5]   A. Prosperetti, Bubbles, Phys. Fluids 16 (2004) 1852–1865. https://doi.org/10.1063/1.1695308

[6]   C. E. Brennen, Cavitation and bubble dynamics, Cambridge University Press, 2013. https://doi.org/10.1017/CBO9781107338760

[7]   B. Verhaagen, Root Canal Cleaning Through Cavitation and Microstreaming, University of Twente, thesis, 2012.

[8]   E-A. Brujan, K. Nahen, P. Schmidt, A. Vogel, Dynamics of laser-induced cavitation bubbles near an elastic boundary, J. Fluid Mechan. 433 (2001) 251–281. https://doi.org/10.1017/S0022112000003347

[9]   TG. Leighton, The acoustic bubble, Academic, London, 1994.

[10]   A. Henglein, Chemical effects of continuous and pulsed ultrasound in aqueous solutions, Ultrason. Sonochem. 2 (1995) S115–S121. https://doi.org/10.1016/1350-4177(95)00022-X

[11]   M. Versluis, B. Schmitz, A. Von der Heydt, D. Lohse, How snapping shrimp snap: through cavitating bubbles, Science 289 (2000) 2114–2117. https://doi.org/10.1126/science.289.5487.2114

[12]   D. Krefting, R. Mettin, W. Lauterborn, High-speed observation of acoustic cavitation erosion in multibubble systems, Ultrason. Sonochem. 11 (2004) 119–123. https://doi.org/10.1016/j.ultsonch.2004.01.006

[13]   W. J. C. Terwisga, P. A. Fitzsimmons, L. Ziru, E. J. Foeth, Cavitation erosion – a review of physical mechanisms and erosion risk models, Proc. 7th Internat. Symp. Cavitation, CAV2009, 16-20 August 2009, Michigan, USA.

[14]   T. J. Mason, D. Peters, Practical sonochemistry: Power ultrasound uses and applications, 2nd edition, Horwood Chemical Science Series, Elsevier, 2002.

[15]   A.D. Walmsley, S.C. Lea, F. Bernhard, D.C. King, G.J. Price, Mapping cavitation activity around dental ultrasonic tips, Clin. Oral Invest. 17 (2013) 1227–1234. https://doi.org/10.1007/s00784-012-0802-5

[16]   A. Walmsley, S. Lea, G. Landini, A. Moses, Advances in power driven pocket/root instrumentation, J. Clin. Periodontol. 35 (2008) 22-28. https://doi.org/10.1111/j.1600-051X.2008.01258.x

[17]   D. Fernandez Rivas, B. Verhaagen, J. Seddon, A. Zijlstra, L.-M. Jiang, L. van der Sluis, M. Versluis, D. Lohse, H. Gardeniers, Localized removal of layers of metal, polymer, or biomaterial by using ultrasound cavitation bubbles, Biomicrofluidics 6 (2012) 34114. https://doi.org/10.1063/1.4747166

[18]   S.C. Lea, B. Felver, G. Landini, A.D. Walmsley, Three-dimensional analyses of ultrasonic scaler oscillations, J. Clin. Periodont. 36 (2009) 44–50. https://doi.org/10.1111/j.1600-051X.2008.01339.x

[19] S.C. Lea, B. Felver, G. Landini, A.D. Walmsley, Ultrasonic scaler probe oscillations and tooth surface defects, J. Dent. Res. 88 (2009) 229–234. https://doi.org/10.1177/0022034508330267

[20] S.C. Lea, G. Landini, A.D. Walmsley, Vibration characteristics of ultrasonic scalers assessed with scanning laser vibrometry, J. Dent. 30 (2002) 147–151. https://doi.org/10.1016/S0300-5712(02)00009-X

[21] S.C. Lea, A.D. Walmsley, Mechano-physical and biophysical properties of power-driven scalers: driving the future of powered instrument design and evaluation, Periodont. 2000 51 (2009) 63-78. https://doi.org/10.1111/j.1600-0757.2009.00300.x

[22] J. West, J. Roane, A. Goerig, Cleaning and shaping the root canal system, in: Cohen S, Burns RC (Eds.), Pathways of the pulp, 6th ed., St. Louis, CV Mosby Co., 1994, pp. 179–81.

[23] FS. Weine, Endodontic therapy, 5th ed., St. Louis, CV Mosby Co., 1996, p. 305.

[24] K. Gulabivala, B. Patel, G. Evans, Y.-L. Ng, Effects of mechanical and chemical procedures on root canal surfaces, Endod. Topics 10 (2005) 103–122. https://doi.org/10.1111/j.1601-1546.2005.00133.x

[25] L. van der Sluis, M. Versluis, M. Wu, P. Wesselink, Passive ultrasonic irrigation of the root canal: a review of the literature, Int. Endod. J. 40 (2007) 415-426. https://doi.org/10.1111/j.1365-2591.2007.01243.x

[26] M. Haapasalo, U. Endal, H. Zandi, J. Coil, Eradication of endodontic infection by instrumentation and irrigation solutions, Endod. Topics 10 (2005) 77-102. https://doi.org/10.1111/j.1601-1546.2005.00135.x

[27] H.-C. Flemming, J. Wingender, U. Szewzyk, Biofilm Highlights, 1st ed., J. W. Costerton (Ed.), Springer Series on Biofilms, Springer-Verlag, Berlin, 2011. [28] R. Macedo, J. Robinson, B. Verhaagen, A. Walmsley, M. Versluis, P. Cooper, L. van der Sluis, A novel methodoly providing insights into removal of biofilm-mimicking hydrogel from lateral morphological features of the root canal during irrigation procedures, Int. Endod. J. 47 (2014) 1040-1051. https://doi.org/10.1111/iej.12246

[29] W. M. Dunne Jr., Bacterial adhesion: seen any good biofilms lately?, Clin. Microbiol. Rev. 15 (2002) 155-166. [30]G. Sagvolden, I. Giaever, J. Feder, Characteristic Protein Adhesion Forces on Glass and Polystyrene Substrates by Atomic Force Microscopy, Langmuir 14 (1998) 5984-5987. https://doi.org/10.1128/CMR.15.2.155-166.2002

[31] B. Thilo, P. Baehni, Effect of ultrasonic instrumentation on dental plaque microflora in vitro, J. Period. Res. 22 (1987) 518-521. https://doi.org/10.1111/j.1600-0765.1987.tb02063.x

[32] J. Thacker, An approach to the mechanism of killing cells in suspension by ultrasound, Biochem. Biophys. Acta 304 (1973) 240-248. https://doi.org/10.1016/0304-4165(73)90241-9

[33] AD. Walmsley, W. Laird, A. Williams, A model system to demonstrate the role of cavitational activity in ultrasonic scaling, J. Dent. Res. 63 (1984) 1162-1165. https://doi.org/10.1177/00220345840630091401

[34] AD. Walmsley, W. Laird, A. Williams, Dental plaque removal by cavitational activity during ultrasonic scaling, J. Clin. Periodont. 15 (1988) 539-543. https://doi.org/10.1111/j.1600-051X.1988.tb02126.x

[35] AD. Walmsley, T. Walsh, W. Laird, et al., Effects of cavitational activity on the root surface of the teeth during ultrasonic scaling, J. Clin. Periodont. 17 (1990) 306-312 https://doi.org/10.1111/j.1600-051X.1990.tb01094.x

[36] K. Bless, B. Sener, J. Dual, T. Attin, P. Schmidlin, Cleaning ability and induced dentin loss of a magnetostrictive ultrasonic instrument at different power settings, Clin. Oral Investig. 15 (2011) 241–248. https://doi.org/10.1007/s00784-009-0379-9

[37] T. Flemmig, G. Petersilka, A. Mehl, R. Hickel, B. Klaiber, The effect of working parameters on root substance removal using a piezoelectric ultrasonic scaler in vitro, J. Clin. Periodont. 25 (1998) 158–163. https://doi.org/10.1111/j.1600-051X.1998.tb02422.x

[38] T. Thurnheer, E. Rohrer, G. Belibasakis, T. Attin, P. Schmidlin, Static biofilm removal around ultrasonic tips in vitro, Clin. Oral Investig. 1 (2014) 1779-1784 . https://doi.org/10.1007/s00784-013-1157-2

[39]   J-B. Park, Y. Jang, M. Koh, B-K. Choi, K-K. Kim, Y. Ko, In vitro analysis of the efficacy of ultrasonic scalers and a toothbrush for removing bacteria from resorbable blast material titanium disks, J. Periodont. 84 (2013) 1191-1198. https://doi.org/10.1902/jop.2012.120369

[40]   WS. Rasband, ImageJ, U. S. National Institutes of Health, Bethesda, Maryland, USA, http://imagej.nih.gov/ij/ (last accessed January 2017)

[41]   http://fiji.sc/Auto_Threshold (last accessed January 2017)

[42]   J. Schindelin, I. Arganda-Carreras, E. Frise, V. Kaynig, et al., Fiji: an open-source platform for biological-image analysis, Nature Meth. 9 (2012) 676-682. https://doi.org/10.1038/nmeth.2019

[43]   H. McMurray, B. Wilson, Mechanistic and spatial study of ultrasonically induced luminol chemiluminescence, J. Phys. Chem. A 103 (1999) 3955-3962. https://doi.org/10.1021/jp984503r

[44]   G. Price, T. Tiong, D. King, Sonochemical characterisation of ultrasonic dental descalers, Ultrason. Sonochem. 21 (2014) 2052–2060. https://doi.org/10.1016/j.ultsonch.2013.12.029

[45]   L-M. Jiang, B. Verhaagen, M. Versluis, L. van der Sluis, Evaluation of a sonic device designed to activate irrigant in the root canal, J. Endodon. 36 (2010) 143–146. https://doi.org/10.1016/j.joen.2009.06.009

[46]   H. Peeters, R. De Moor, D. Suharto, Visualization of removal of trapped air from the apical region in simulated root canals by laser-activated irrigation using an Er, Cr: YSGG laser, Lasers Med. Sci. 30 (2015) 1683-1688. https://doi.org/10.1007/s10103-014-1643-z

[47]   E. Pecheva, RL. Sammons, AD. Walmsley, The performance characteristics of a piezoelectric ultrasonic dental scaler, Med. Engin. Phys. 38 (2016) 199-203. https://doi.org/10.1016/j.medengphy.2015.10.008

[48]   R. Maurus, V. Ilchenko, T. Sattelmayer, Automated high-speed video analysis of the bubble dynamics in subcooled flow boiling, Intern. J. Heat Fluid Flow 25 (2004) 149–158. https://doi.org/10.1016/j.ijheatfluidflow.2003.11.007

[49]   H. Chen, X. Li, M. Wan, S. Wang, High-speed observation of cavitation bubble cloud structures in the focal region of a 1.2 MHz high-intensity focused ultrasound

transducer, Ultrason. Sonochem. 14 (2007) 291–297.
https://doi.org/10.1016/j.ultsonch.2006.08.003

[50]   G. Duhar, C. Colin, Dynamics of bubble growth and detachment in a viscous shear
flow, Phys. Fluids 18 (2006) 077101. https://doi.org/10.1063/1.2213638

[51]   R. Macedo, B. Verhaagen, D. Fernandez Rivas, J. Gardeniers, L. van der Sluis,
Wesselink, et al., Sonochemical and high-speed optical characterization of
cavitation generated by an ultrasonically oscillating dental file in root canal
models, Ultrason. Sonochem. 21 (2014) 324–335.
https://doi.org/10.1016/j.ultsonch.2013.03.001

[52]   N. Vyas, E. Pecheva, H. Dehghani, RL. Sammons, Q. Wang, D. Leppinen, AD.
Walmsley, High speed imaging of cavitation around dental ultrasonic scaler tips,
PLoS ONE 11 (2016) e0149804. https://doi.org/10.1371/journal.pone.0149804

[53]   B. Felver, D. King, S. Lea, G. Price, AD. Walmsley, Cavitation occurrence around
ultrasonic dental scalers, Ultrason. Sonochem. 16 (2009) 692–697
https://doi.org/10.1016/j.ultsonch.2008.11.002

[54]   N. Kuznetsov, V. Mazya, B. Vainberg, Linear water waves: a mathematical
approach, Cambridge University Press, Cambridge, UK, 2002.
https://doi.org/10.1017/CBO9780511546778

CHAPTER 7

Cleaning Efficacy of Commercial Toothpastes and Toothbrushes

Abstract

Oral health is a determinant factor for the quality of life, essential for well-being and an integral part of general health. Oral diseases affect all age groups and are associated with a high cost of care. Dental caries is one of the most common oral diseases and if left untreated at the early stage can progress to cavities. Worldwide, 60–90% of school children and nearly 100% of adults have dental cavities, often leading to pain and discomfort. Dental cavities and periodontal disease are the major causes of tooth loss. Globally, about 30% of people aged 65–74 have no natural teeth [1].

Oral health is a state of being free from mouth and facial pain, oral infection and sores, periodontal (gum) disease, tooth decay, tooth loss, and other diseases and disorders that limit an individual's capacity in biting, chewing, smiling, speaking, and psychosocial wellbeing. Oral diseases affect all age groups and are associated with a high cost of care. In the EU, the socio-economic burden of oral diseases is considerable: they affect the majority of school-aged children and adults and account for 5% of public health spending. Costs of traditional curative treatment have risen from €54 bn in 2000 to €79 bn in 2012 and are expected to rise up to €93 bn by 2020 [2]. Treatment expenditure exceeds that for other diseases, including cancer, heart disease, stroke and dementia. This is disturbing, given that much of the oral disease burden is preventable.

The most common oral diseases are dental cavities, periodontal (gum) disease, oral cancer, oral infectious diseases, trauma from injuries, and hereditary lesions. Dental caries is a disease of the hard tissues of the teeth caused by the interactions over time between micro-organisms found in dental plaque and dietary fermentable carbohydrates (principally sugars such as sucrose). Dental decay is easily preventable, but is nonetheless one of the most common chronic diseases. Worldwide, 60–90% of school children and nearly 100% of adults have dental cavities, often leading to pain and discomfort. Dental caries if left untreated at the early stage can progress to cavities. Dental cavities and periodontal disease are the major causes of tooth loss. Complete loss of natural teeth is widespread and particularly affects older people. Globally, about 30% of people aged 65–

74 have no natural teeth. Most of us are not aware that the early stages of damage can be stopped or reversed with changes to diet and personal dental hygiene practices supplemented by fluorides (brushing, mouth washing, flossing, etc.). At a later stage, changes seen by the dentist as either localized breakdown of the surface enamel or an underlying dark shadow from the inner dentine can be controlled by more intensive preventive treatments and homecare, whilst others may require tooth-preserving fillings. At the extensive-stage the dentist sees a distinct physical cavity with visible internal dentine. These lesions are likely to require tooth-preserving filling as well as preventive control of the underlying disease factors [3].

In Europe, the total spend on dental health each year is in the region of €40 billion and around 66% of these costs relate to treating dental caries and its consequences [4]. In Europe, the number of people with dental caries is falling. But 80% of cases are found in just 20% of the population [5]. If healthy diet, regular dental check-ups and routine personal oral hygiene practices do not become a habit for the patient, the next stage is treatment in the dental cabinet.

Keywords

Oral Health, Plaque Cleaning, Toothbrush, Toothpaste, Granules, Tooth Model, Cheek Model

Contents

1.   Cleaning the *In Vitro* Plaque Off Teeth Models

Maintenance of oral health depends on dental plaque removal and prevention of its accumulation. Toothpastes (TPs) with various compositions are used to reduce plaque, calculus, gingivitis, and the onset and progression of periodontitis [6].

There is a need to assess and quantify the efficacy of TPs for plaque removal. Laboratory evaluations of TP effect on plaque control ultimately will require clinical studies to document clinical benefit. However, it is difficult to record clinically the TP interdental plaque removal due to the anatomy of the human jaw. Accurate and reproducible *in vitro* tests are a valuable alternative because they can indicate where clinical research should focus and help to explain *in vivo* results from clinical studies. Laboratory studies on the removal of tooth surface deposits have routinely been used for evaluation of TB efficacy with or without a TP. Artificial plaque deposits have also been developed to enable evaluation on the removal of supra-gingival and sub-gingival plaque [7-9]. Several of these findings have been correlated with clinical indices [10-12]. Specially constructed toothbrushing simulators are used for the *in vitro* tests [13]. Many of them move with a linear motion or they are rotating over a stationary tooth specimen. The use of robotic arms to provide a standardized cleaning mechanism has been used [14]. All simulators aim to orientate and move the filaments to reach all areas of the teeth from all angles, especially in hard to reach positions, such as the interproximal and lingual areas.

Important variable in studying the plaque removal is the load applied during toothbrushing. The loading of TBs has been measured by applying a strain gauge to the neck (between head and handle) of the TB under loading values ranging from 2.74 to 2.91 N. Clinical research has shown that the relationship between brushing force and plaque removal efficacy of manual toothbrushes increases up to a certain level of force at around 2.94 N [15]. There is reduced efficiency beyond this force: because of the high force applied the filaments are splayed and may become less effective in removing plaque.

Not at a last place, the formulation of the TPs is influential in assuring efficient cleaning of the dental plaque. Small chemical additions such as zinc (Zn) or silver (Ag) provide an antibacterial effect, whilst calcium (Ca) and fluoride (F) helps remineralization. On the other hand, the incorporation of submicron particles (granules) physically helps to remove plaque and assists interdental cleaning. Combined chemical and mechanical action of the TPs can have an synergistic effect in fighting plaque accumulation and this will aid in achieving better oral health.

The aim of this work was to study and quantify plaque removal efficacy from interdental areas of a Zn-based TP containing silica microgranules. The study was expected to assess whether the microgranules aid in achieving better teeth cleaning. Three sample groups were prepared for:

I. Brushing with the TP, containing silica microgranules (SiO_2-Zn-TP).
II. Brushing with the TP without silica microgranules (Zn-TP).

III. Brushing with a toothbrush without TP, only with water as control.

Tooth models were prepared with natural teeth using two lower molars (6th and 7th) and one lower premolar (Fig. 77 a). *S. mutans* biofilm was used as a model *in*-vitro plaque in this work as one of the most prevalent species in human plaque that is strongly associated with caries production in teeth. The biofilm growth was described in section 2 of chapter 5. A digital camera Nikon D40 was used to take general pictures of the tooth models before and after the toothbrushing. The LM systems used for imaging the interdental spaces of the tooth models in more details were Nicon Eclipse TE300 (Japan) and GE-5 digital microscope (View Solutions Inc., USA): image analyses were based on these images. SEM (Zeiss EVO MA-10, Germany) was used to obtain better magnification of the biofilm in the interdental spaces. Models were measured at 5 kV without gold sputtering. Sputtering with gold nanocoating used to obtain high quality images was not possible due to the high moisture contained in the acrylic in which the teeth were embedded which hampered also the measurement and there was a lot of charging as seen in the SEM images (white areas which do not have to be associated with plaque features).

A robot (Kuka Industial, Roboter GmbH, Germany) was used to provide a repeatable and precise brushing action simulating a clinical movement of the TB. Each model was fixed in a brushing rig (Fig. 99 a) and TP pea sized amount as recommended by dentists was weighed for each measurement (≈ 0.5g) [16]. A moistened TB with a TP was fixed in the robot arm (Fig. 99 b) and moved over the facial surface of the tooth model with an oscillating-rotation motion at 0.8 Hz for 12 sec, equivalent to 10 brush cycles and under load of 2.5 N. The load was measured by a precise digital balance (Timetop, capacity 1000 g, precision 0.1 g). Since the TB was mechanically fixed to the robot arm, which reduces the elasticity of its handle in comparison to when a hand is holding the TB, higher loads were considered to be not appropriate due to the risk of deformation and breaking of the TB during brushing. It was also observed that at higher loads the filaments of the TB were strongly bent. Clinical research has shown that the relationship between brushing force and plaque removal efficacy of manual toothbrushes increases up to a certain level of force at around 2.94 N and beyond this force the efficiency is reduced [15]. Because of the applied force, the filaments are splayed and may become less effective in removing plaque. Brushing with more than 3 N may also cause gingival recession [17]. Other studies reveal optimal brushing forces ranging from 2.3 N to 3.23 N [17]. The discrepancies might have been due to random effects such as using different measurement systems, toothbrushes, gender, age and dental characteristics of the study groups. In terms of the efficacy of the brushing duration, dentists recommend brushing for at least 2 min of all 32 teeth [16]. Since our tooth model had 3 teeth this was equivalent to approximately 12 sec brushing time.

(a) brushing rig   (b) robot set-up

Figure 99. Experimental set-up used for the brushing experiment.

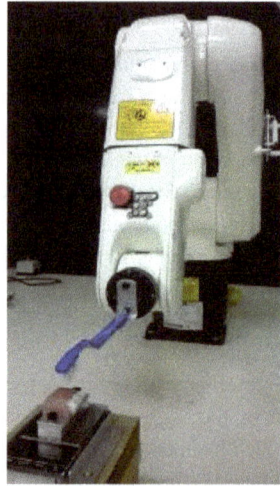

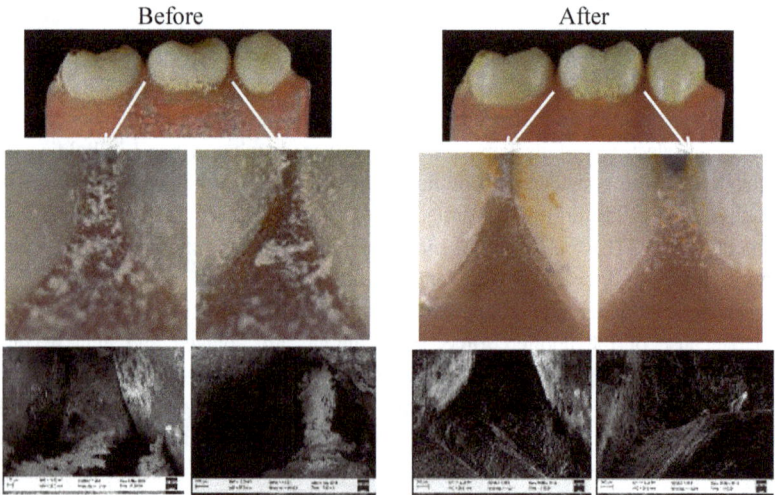

Figure 100. Before and after images of a model brushed off with SiO_2-Zn-TP, taken with a digital camera (top), LM (middle) and SEM (down).

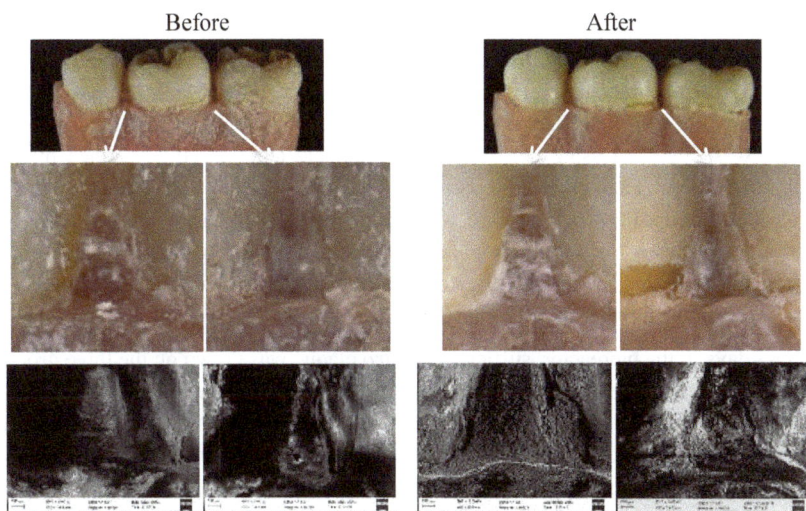

Figure 101. Before and after images of a model brushed off with Zn-TP, taken with a digital camera (top), LM (middle) and SEM (down).

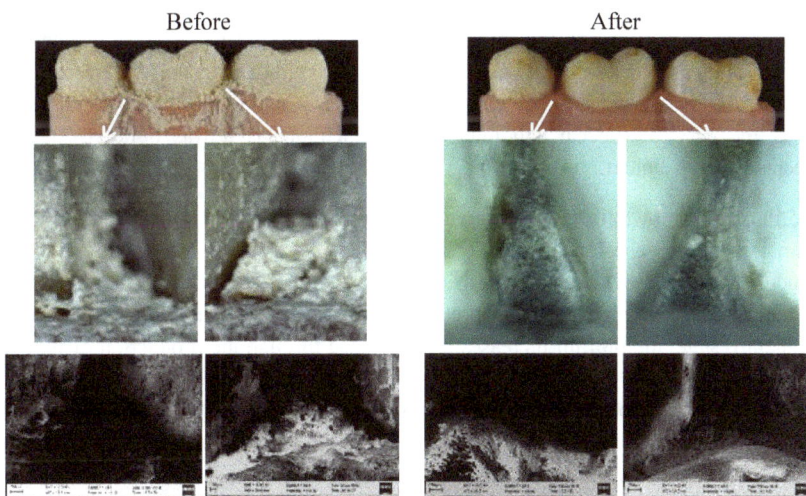

Figure 102. Before and after images of a model brushed off with water, taken with a digital camera (top), LM (middle) and SEM (down).

After the toothbrushing, running distilled water from a lab bottle was used to remove remaining TP and remaining plaque debris before imaging. The modles were left to dry in air at room temperature. After staining the *in-vitro* plaque with plaque disclosing solution for 60 sec (PlaqSearch™, TePe Munhygienprodukter, Sweden), the remaining plaque was estimated using the Quigley Hein plaque index (QHPI), defined by G. A. Quigley and J. W. Hein [18] and modified by S. Turesky [19]. Three dentists with strong clinical and scientific experience were asked for their "blind" estimation, i.e. they were given the images of all tooth models and they did not know which group the models belong to. An index for the three teeth in each model was determined by dividing the total score by the number of teeth examined.

After 3 days in the bioreactor, sufficient *in-vitro* plaque was grown on the three groups of natural tooth models. Six models of each group, i.e. to be brushed off with the SiO$_2$-Zn-TP (Fig. 100), with the Zn-TP (Fig. 101) and with no TP, just water as control (Fig. 102) were prepared. One of the models of each group was used for SEM observation (Fig. 103). The biofilm was imaged with a digital camera and more details of the interdental spaces of interest were obtained with LM and SEM as revealed in Figs. 100-102. Left and right interdental spaces were observed and used in the image analysis for each model. For comparative reasons SEM control images of the tooth and the acrylic surfaces without grown plaque were taken. As can be seen in Fig. 103, the tooth surface is relatively smooth with nothing grown on it, while the acrylic has specific relief structure. When the *in-vitro* plaque was grown, the smooth tooth surface and the acrylic were hidden behind spongy plaque layer with flakes. However, after toothbrushing these were completely brushed off the acrylic but part of the spongy layer remained on the tooth surface and along the cervical (buccal and interstitial), which means it was quite resistant to a mechanical detachment.

Biofilms are known to adhere strongly to a substrate. The initial state of biofilm formation is the adhesion of planktonic bacteria to a substrate, after which the biofilm structure is developed. The bacteria themselves adhere to the substrate due to a multitude of forces, based on physicochemical and molecular/cellular interactions [20,21]. During its mature state, the biofilm increases its adherence through the production of functional amyloids, among others [22]. Properties of the substrate itself are important for the biofilm adhesion, especially the surface roughness and hydrophobicity [21,22] and in this work the natural teeth and the PMMA resin are actually quite rough. Clinical studies reported in the literature are based on brushing of real oral plaque while most of the laboratory studies report work with wet *in-vitro* plaque containing one or several species of live bacteria [23]. Comparison of various wet and dehydrated biofilms reveals that the dehydration changes the biofilm morphology, which is expressed in structure shrinkage when the water is removed from the cell wall [24]. On the other hand, studies also show change of the nanomechanical and electrical properties of biofilms upon dehydration [25]. However there is no data on how the dehydration influences the biofilms adhesive properties.

Tooth with no plaque: Acrylic with no plaque:

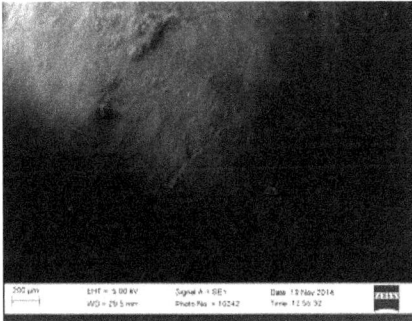

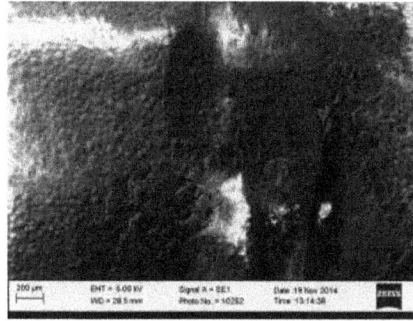

Before brushing of plaque: After brushing with a TP:

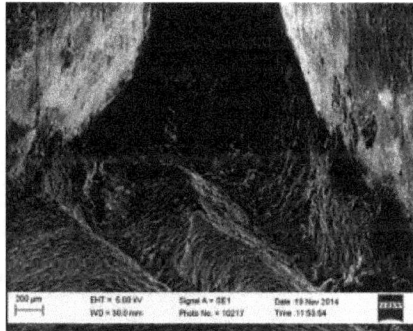

Figure 103. SEM images of the tooth surface and PMMA acrylic with no in-vitro plaque grown (upper row). Images of a model brushed off with SiO$_2$-Zn-TP (down row).

Qualitative assessment of the remaining plaque was performed by three experienced dentists and the results are shown graphically in Fig. 104. The results were determined from the scores of the stained plaque on the facial tooth surfaces using the QHPI [19,19]. Dental plaque was assessed after brushing the teeth in the model and the mean plaque index was then assessed qualitatively (Table 11). Both the figure and the table show how inconsistent this type of estimation can be since the three dentists had different scores for the three groups of models. The assessment was even more difficult because it differs from the clinical estimation where the oral plaque is formed along the cervical of the tooth whilst in our study it may cover the entire tooth surface, which will produce a difference in the results. The overall interdental score showed more plaque remaining after brushing with the SiO_2-Zn-TP versus the Zn-TP. On the other side brushing with both SiO_2-Zn-TP and Zn-TP was more efficient than brushing with water only. It has to be pointed out that the QHPI is only a qualitative estimation of the TPs brushing efficacy.

Table 11. Overall interdental scores for in-vitro plaque remaining after toothbrushing.

Group	Mean QHPI ($\pm$ SD)
SiO_2-Zn-TP	2.0 ± 0.9
Zn-TP	1.8 ± 0.6
Water	2.2 ± 1.0

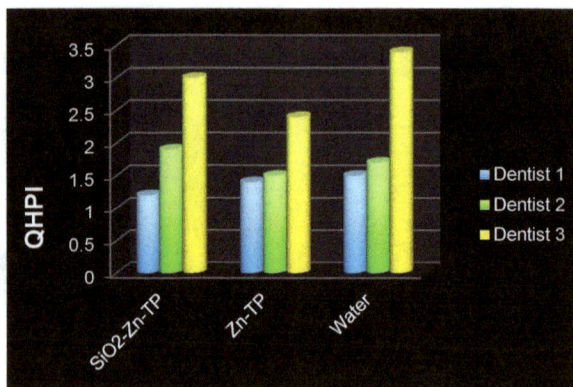

Figure 104. QHPI estimation of in-vitro plaque remaining after toothbrushing.

Using the LM images of the tooth models it was possible to estimate quantitatively the plaque removal from the interdental spaces. A triangular area with the same size was chosen for the analysis to best fit the shape of the interdental space and to ensure repeatability of the results for every tooth model and each interdental space (Fig. 105).

Data for the five models in each group was estimated in pixels using a threshold method on an 8-bit smoothed image in ImageJ software for image analysis, which shows the biofilm as white pixels and the non-covered area as black pixels [26-29]. Pixels representing the plaque before brushing for the five models in each group were averaged and this pixel number was taken for 100% of grown plaque. Pixels for the plaque remaining after brushing were averaged too and the percentage of removed plaque was calculated.

Figure 105. A triangular area chosen for the image analysis of the interdental spaces.

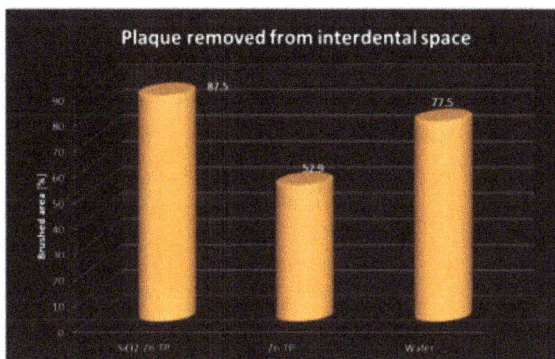

Figure 106. Image analysis of the plaque removed from interdental spaces after toothbrushing.

Statistical differences for the removed plaque were analysed with IBM SPSS statistical software (v.21 for Windows, SPSS Inc., Chicago, IL, USA), using 1-way ANOVA and multiple post hoc comparisons (Tukey test) at a significance level of $p < 0.05$, with the dependent variable being the removed plaque.

Image analysis was used to quantify the plaque removal efficacy of the SiO_2 microgranules contained in the SiO_2-Zn-TP after brushing the interdental spaces of the tooth models. This TP effect was compared to a Zn-containing only TP (Zn-TP). Toothbrushing only with water to extract the effect of the TB was carried out. Figure 106 shows the percentage of removed plaque for the three groups of models. Significant plaque removal ($p = 0.001$) was obtained when brushing with the SiO_2-Zn-TP (87.5%) versus brushing with Zn-TP (52.9%). The presence of SiO_2 microgranules in the TP aids in achieving better teeth cleaning. The SiO_2-Zn-TP provided better efficacy in the plaque removal (87.5%) than the water brushing (77.5%) which can be attributed to a facilitated mechanical action with the SiO_2 microgranules but the difference is not statistically significant (p = 0.395). However, the efficacy of the Zn-TP is significantly lower (p = 0.016) compared to the effect of the TB alone (i.e. when brushing with water). It is known that brushing with a TB relies on the mechanical action of the filaments. It is likely that the Zn-TP diminishes the effect of mechanical cleaning of the TB by "lubricating" the filaments. There is no conclusive evidence in the literature for the precise role of dentifrice and its contribution to plaque removal during mechanical toothbrushing. For instance, one study [30] showed that brushing with a dentifrice removed more plaque than brushing without one; another found no difference between brushing with or without a dentifrice [31], and two others found more plaque reduction when no dentifrice was used [32,33]. However Paraskevas and Forward demonstrated that toothbrushing alone is sufficient to remove plaque and that the addition of a TP is unnecessary in removing it [32,34], although the use of a TP has been shown to reduce the rate of plaque growth after brushing, compared to brushing with water only [35]. These inconsistencies can be attributed to factors related to study methodology (differences in the choice of indices for plaque assessment), the treatment provided (brushing time, professional versus manual toothbrushing and rinsing prior to plaque assessment), study population (children versus adults) and study design (split-mouth versus whole mouth). On the other hand TPs, which contain only a chemical element such as fluoride, calcium phosphate, zinc, chloride, etc. aim mainly at improving teeth antibacterial resistance, mineralization/remineralization rates, or reducing enamel erosion and abrasion [17]. A reason for the lower efficiency of the Zn-TP versus the TB alone can be the absence of a mechanical element such as the SiO_2 microgranules.

2. *In Vitro* Study of Toothpaste Efficacy to Remove Interdental Plaque - Summary

In this study, the plaque removal efficacy of a commercial TP with silica microgranules was quantified by image analysis instead of using the more popular clinical method of indexing the remaining plaque and estimation by the QHPI which gives only qualitative results and is ambiguous. It was found that brushing with the microgranules was superior to brushing with Zn-containing only TP. This means that the microgranules aid the teeth cleaning. The cleaning efficacy of the Zn-containing TP was lower compared to brushing with water only. We assume that brushing with a TB relies on the mechanical action of the filaments and is superior to brushing with TP containing chemical elements only and no abrasive particles.

3.   Cleaning *In Vitro* Biofilm off Cheek Models

There is an accepted correlation between poor oral hygiene due to the accumulation of oral plaque and oral health conditions such as caries, gum disease and malodour [36]. Hence, mechanical removal of plaque by toothbrushing, which should take place on a regular basis twice a day, has long been considered the foundation of good oral hygiene practice [37]. Successful toothbrushing also depends on motivation of the individual, their knowledge on what the procedure is aiming to achieve and the manual dexterity of the patient. To improve the efficient removal of oral biofilms and their associated bacteria, TB design continues to evolve. The design of the TB is important as brushing is more effective with a small-headed toothbrush with soft, round-ended filaments, a compact, angled arrangement of long and short filaments and a handle, which is comfortable [38]. TB features can now include angled heads, tongue cleaners and a multitude of different tufting patterns. These features aid the user to complete a thorough clean of their whole mouth. The TB design is important in order to apply the correct amount of pressure (pressure = force / area) to the filaments as they contact the tooth. Previous research has shown that an angled neck leads to superior plaque removal due to a better access around the mouth (easier reach of back teeth), a correct amount of pressure being applied which yields a superior plaque removal [39]. There are several dual-face TBs on the market nowadays which allow to brush the teeth and the tongue/cheek at the same time.

The aim of this work was to assess the ability of commercial dual-face TB to clean bacterial plaque from teeth and cheek simultaneously. The TB filaments were arranged in alternating arrays of transverse filaments and wide-angle fan filaments. The back of the TB head was covered in an arrangement of soft rubber scrapers. The brush was designed to simultaneously clean both teeth and cheek to reduce unpleasant odour. We selected as a positive control a standard TB (Colgate 360). Four sample groups have been prepared for:

I.   Brushing the cheek models with the tested dual-face TB.
II. Brushing the tooth models with the tested dual-face TB.
III. Brushing the cheek models with Colgate 360 TB as a control.
IV. Brushing the tooth models with Colgate 360 TB as a control.

Tooth models were prepared with natural teeth embedded in a PMMA matrix and cheek models were prepared from sandblasted spherical vinyl sheets, as described in details in chapter 5 (Figs. 77 and 79, sections 3 and 4 respectively). *S. Mutans* was used as a model *in*-vitro plaque on the tooth models and two-species biofilm of *S. Sanguinis* and *S.*

Mutans was chosen as a model *biofilm* on the cheek models. Details for the bacteria growth on the two models were given in chapter 5.

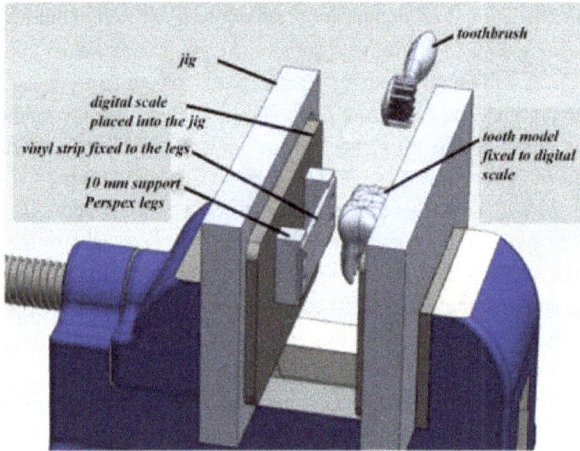

Figure 107. Eexperimental set-up for the brushing experiment.

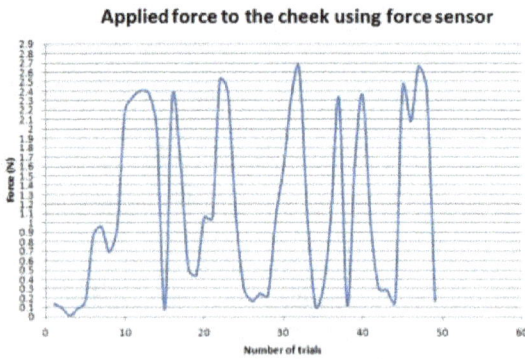

Figure 108. Graph showing the readings from the sensor while volunteers were brushing their cheek.

A robot (Kuka Industial, Roboter GmbH, Germany, Fig. 99 b) was used to provide a repeatable and precise brushing action simulating a clinical movement of the TB. The brushing rig (Fig. 107) consisted of a large engineering vice with attached fine digital scales (Timetop, capacity 1000 g, precision 0.1 g) on the two jaws. Cheek model was then fixed on one of the scales and the tooth model on the other one. A moistened TB was fixed in the robot arm and placed between the two models. The sliding of the jaws allowed us to approach the two models towards each other in small steps until the TB exerted a load on the two models. The load on the tooth model was fixed at 2.5 N as explained in section 1. The load on the cheek model was set after an experiment with three volunteers which brushed their cheek with a standard TB that had a fine force sensor (FlexiForce, Tekscan, USA), attached to the back of the TB head and allowing measurements in humid environments. The fifty readings were read and analysed by a home-written programme, which gave an average value of 1.1 N (Fig. 108). For the brushing experiment, the TB was moved along the facial surface of the tooth model with an oscillating-rotation motion at 0.8 Hz for 12 sec, equivalent to 10 brush cycles. After the end of toothbrushing, running distilled water from a lab bottle was used to remove remaining plaque debris from the models before imaging. The models were left to dry in air at room temperature.

The biofilm on both models was imaged with a digital camera and more details were obtained with LM and SEM as revealed in Figs. 109-112. The cheek models were covered with dense white biofilm and the LM images showed white clusters of bacteria. SEM also showed dense bacterial population all over the vinyl strips. After brushing with the back of the tested dual-face and control TBs, large area of biofilm was removed from the cheek models. The area cleaned with the tested TB was well depicted (Fig. 109: digital camera image after), while the Colgate 360 TB did not remove the biofilm completely but mainly left scratches (Fig. 110: digital camera image after). LM and SEM images after the brushing (Figs. 109, 110) revealed that most of the biofilm in the brushed area was gone and only debris remained on the vinyl strips.

BEFORE

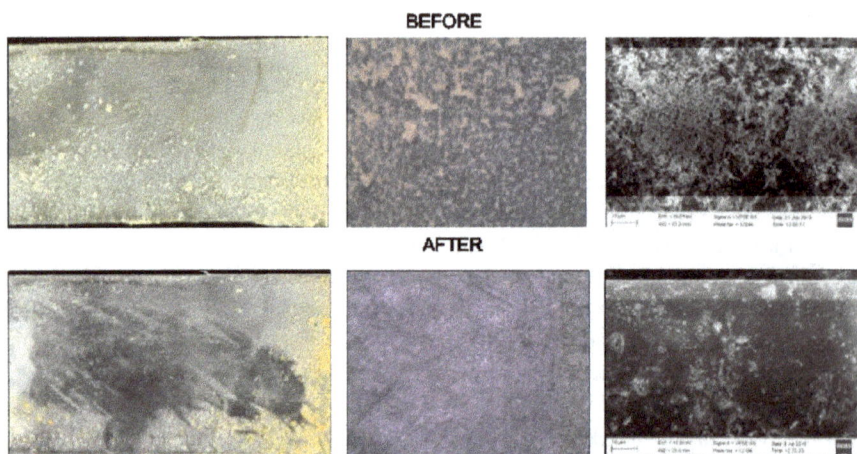

AFTER

Figure 109. Before and after images of a cheek model brushed off with the tested dual-face TB, taken with a digital camera (left), LM (middle) and SEM (right).

BEFORE

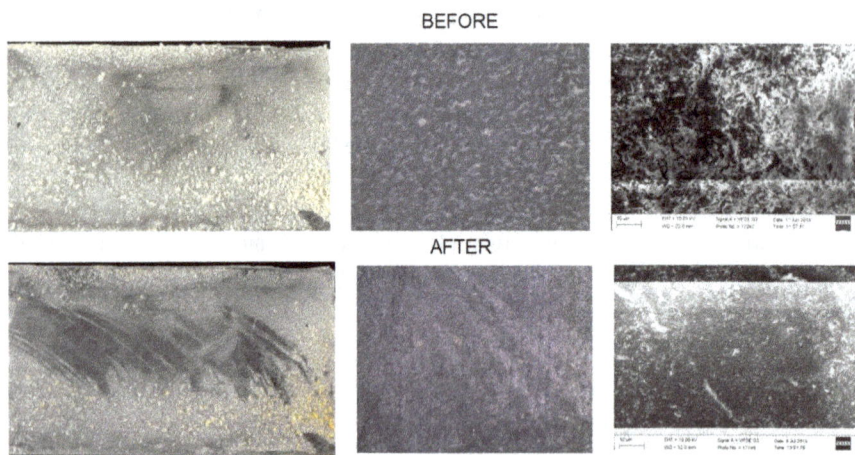

AFTER

Figure 110. Before and after images of a cheek model brushed off with Colgate 360 TB, taken with a digital camera (left), LM (middle) and SEM (right).

BEFORE

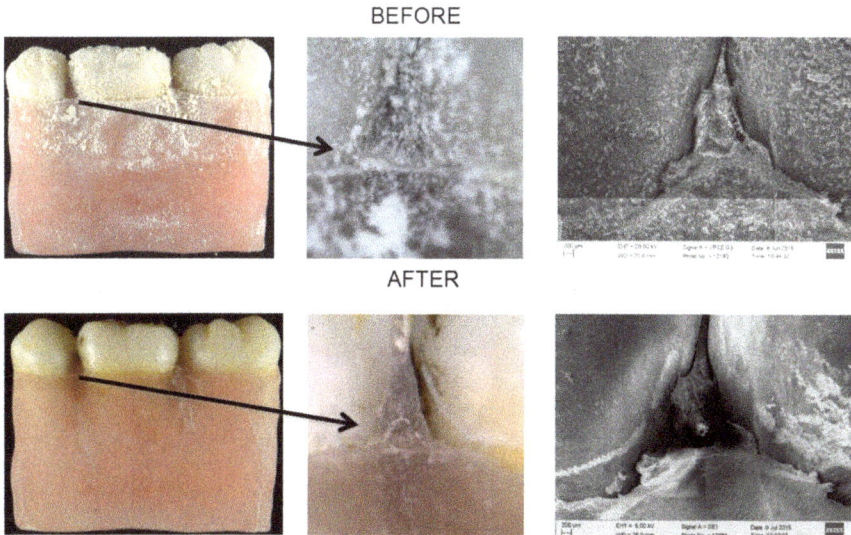

AFTER

Figure 111. Before and after images of a tooth model brushed off with the dual-face TB, taken with a digital camera (left), LM (middle) and SEM (right).

For the tooth models, the interdental spaces were of particular interest. Left and right interdental spaces were observed and used in the image analysis for each model. Digital camera images before (Fig. 111) showed that biofilm was grown all over the tooth models and especially in the interdental spaces as also seen in the LM images in the same figure. SEM showed that the smooth tooth surface and the acrylic gum were hidden behind spongy plaque layer (Figs. 111, 112: SEM images before). After brushing, SEM and LM showed that the *in vitro* plaque was mostly brushed off the teeth and the acrylic gum, with only tiny groups of bacteria left in the interdental spaces (Figs. 111, 112: images after).

Quantitative estimation of the plaque removal off the two models with the tested and control TBs was performed using the digital camera images. For the analysis of the cheek models, a rectangular area was chosen to best fit the shape of the vinyl strip and to ensure repeatability of the results for every cheek model (Fig. 113). Data for the five models in each group was estimated in pixels using a threshold method on an 8-bit smoothed image in ImageJ as described before. Pixels representing the plaque before brushing for the five models in each group were averaged and this pixel number was taken as 100% of grown plaque. Pixels for the plaque remaining after brushing were averaged too and the percentage of removed plaque was then calculated.

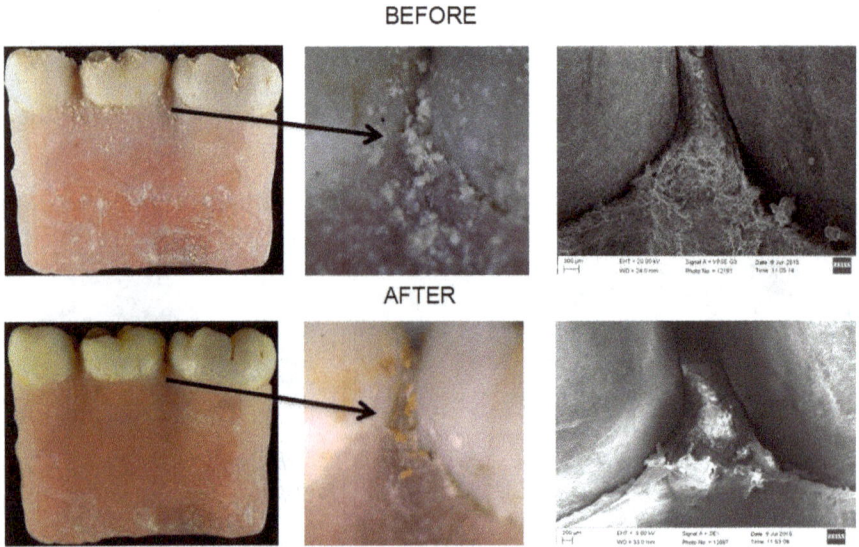

BEFORE

AFTER

Figure 112. Before and after images of a tooth model brushed off with Colgate TB, taken with a digital camera (left), LM (middle) and SEM (right).

Figure 113. Rectangular area chosen for the image analysis of cheek models.

Figure 114. Triangular area chosen for the image analysis of teeth models.

For the tooth models, LM images were used to estimate quantitatively the plaque removal from the interdental spaces. A triangular area was chosen for the analysis to best fit the shape of the interdental space and to ensure repeatability of the results for every tooth model and each interdental space (Fig. 114). Data for the six models in each group was estimated in the way described above for the cheek models.

Statistical differences for the removed plaque were analysed using 1-way ANOVA and multiple post hoc comparisons (Tukey test) with the dependent variable being the removed plaque ($p < 0.05$).

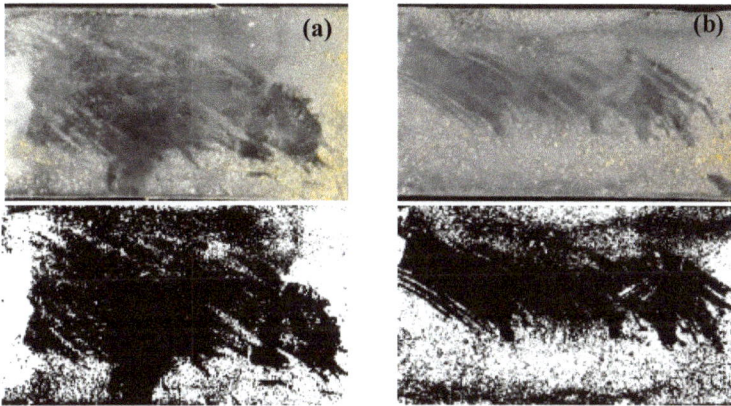

Figure 115. Image analysis of the plaque removed from representative cheek models brushed with the tested dual-face (a) and Colgate 360 (b) TBs.

Image analysis of the plaque removed from the cheek and tooth models are shown in Figs. 115 and 116. Grayscale images of the cheek (Fig. 115) and tooth (Fig. 116) models after the brushing with the tested and control TBs were taken as a basis for obtaining the threshold images (the black and white images in Figs. 115, 116), which were then used to quantify the plaque removal efficacy of the two TBs and the numerical results are presented in Table 12. The tested TB was brushing the cheek models significantly better than the control TB: 30.9% of the plaque was brushed off while Colgate 360 TB removes only 13.9% ($p = 0.03$). The dual-face TB was brushing the teeth as well as the control TB: 89.9% of the plaque in the interdental spaces was cleaned with the tested TB while the control TB removed 92.1% ($p = 0.186$). It was suggested that the longer and more elastic scrapers on the back of the tested TB head were the reason for the better cleaning effect (Fig.117 a), while the shorter and more rigid scrapers of the control TB head mainly scratched the biofilm (Fig. 117 b). It was considered that the more elastic neck of the dual-face TB (Fig. 117 c) facilitated the better cleaning in comparison to the stiff neck of Colgate 360 TB (Fig. 117 d) allowing a better fit to the shape of the underlying model or (in reality) to the curvature of the cheek and the teeth in mouth. No difference in the brushing efficacy of the filaments of the tested TB was found when compared to the control TB filaments. A reason for this result might be the filaments design (Figs. 117 e-h).

Table 12. Quantification of the results from brushing the cheek and tooth models with the tested dual-face and control TBs

	Brushed area ± SD (%)	
	Cheek models	**Tooth models**
Tested dual-face TB	30.9 ± 12.1	89.9 ± 2.4
Control Colgate 360	13.9 ± 1.6	92.1 ± 2.5

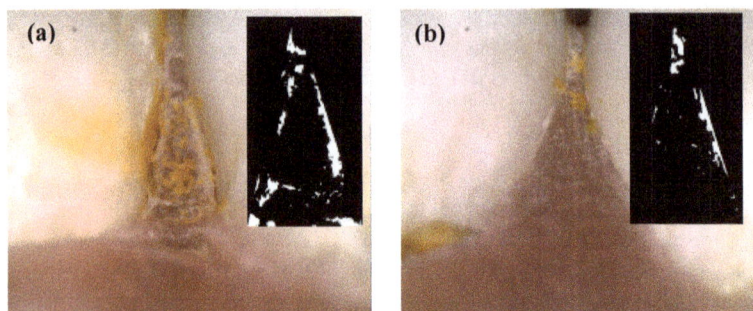

Figure 116. Image analysis of the plaque removed from representative tooth models brushed with the tested dual-face (a) and Colgate 360 (b) TBs.

4. Simultaneous Cleaning of Teeth and Cheeks with a Dual-Face Toothbrush - Summary

A laboratory method showing the efficacy of new angled TB on plaque and mouth odour control was developed and it offers an approach that move *in vitro* testing closer to simulated clinical conditions. The quantification of plaque removal efficacy of TBs no matter of their design was easily achieved using this method and the tooth/cheek surface cleaning potential of the two TBs used was determined. The method has the potential to indicate where the clinical research should focus and will also help to explain *in vivo* findings from clinical trials.

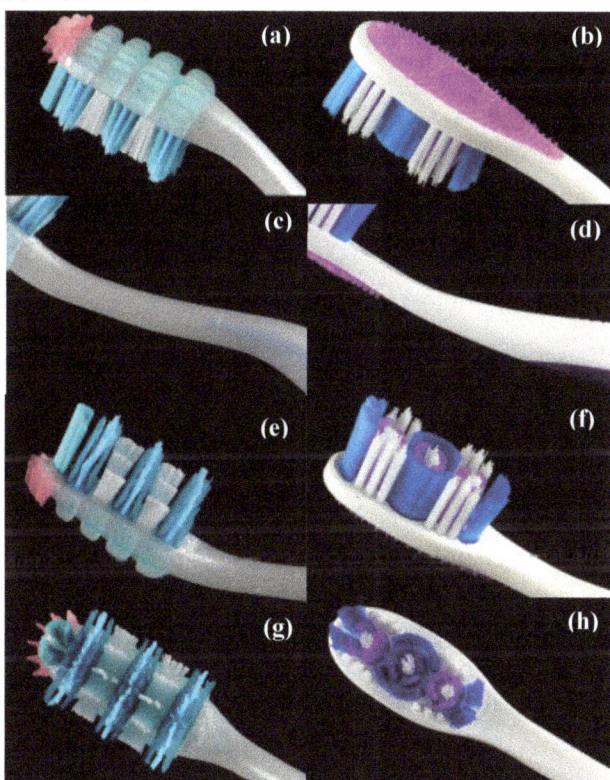

Figure 117. The tested dual-face (left) and Colgate 360 (right) TBs: scrapers on the back of the head for brushing the cheek (a,b); neck (c,d); filaments for brushing the teeth (e-h).

References

[1]   World Health Organization, Report on Oral Health, http://www.who.int/oral_health/media/en/orh_report03_en.pdf (last accessed January 2017).

[2]   Platform for better oral health in Europe, http://www.oralhealthplatform.eu/oral-health/ (last accessed January 2017)

[3]   N. Pitts, K. Ekstrand, The ICDAS Foundation. International Caries Detection and Assessment System (ICDAS) and its International Caries Classification and Management System (ICCMS™) – methods for staging of the caries process and enabling dentists to manage caries, Community Dentistry and Oral Epidemiology 41 (2013) e41-e52.h ttps://doi.org/10.1111/cdoe.12025

[4]   Fluoride and Dental Health in Europe http://www.bfsweb.org/documents/denthlth.PDF (last accessed January 2017)

[5]   E. Widstrom, K. Eaton, Oral healthcare systems in the extended European Union, Oral Health Prev. Dent. 2 (2004) 155-194.

[6]   P. Marsh, M. Martin, Oral Microbiology, 3rd edition, Chapman & Hall, London, 1992h ttps://doi.org/10.1007/978-1-4615-7556-6

[7]   U. Saxer, S. Yankell, A review of laboratory methods to determine toothbrush safety anf efficacy, J. Clin. Dent. 8 (1997) 114-119.

[8]   S. Yankell, X. Shi, R. Emling, Laboratory efficacy of three toothbrushes with soft or medium texture, J. Dent. Res. 80 (2001) 120. [9]   S. Yankell, X. Shi, R. Emling, R. Bucker, S. Loudin, Laboratory evaluation of two bi-level toothbrush products for subgingival access and gingival margin cleaning, J. Clin. Dent. 11 (2000) 20–23.

[10]   A. Volpe, R. Emling, S. Yankell, The toothbrush - A new dimension in design, engineering and clinical evaluation, J. Clin. Dent. 3 (1992) C29–C33.

[11]   S. Singh, M. Deasy, Clinical plaque removal performance of two manual toothbrushes, J. Clin. Dent. 4 (1993) D13-D16.

[12]   A. R. Biesbrock, P. A. Walters, R. D. Bartizek, The relative effectiveness of six powered toothbrushes for dental plaque removal, J. Clin. Dent. 13 (2002) 198-202.

[13]   J. Parry, E. Harrington, G. Rees, R. McNab, A. Smith, Control of brushing variables for the in vitro assessment of toothpaste abrasivity using a novel

laboratory model, J. Dent. 36 (2008) 117–124.h
ttps://doi.org/10.1016/j.jdent.2007.11.004

[14] J. Liu, A novel motion generation strategy for robotic tooth brushing simulator, Ind. Robot 40 (2013) 355–362.h ttps://doi.org/10.1108/01439911311320868

[15] K. Carter, G. Landini, A.D. Walmsley, An in vitro study on the deformation of powered toothbrush filaments, Quintessence Int. 385 (2007) E263-270.

[16] http://www.nhs.uk/Livewell/dentalhealth/Pages/Teethcleaningguide.aspx (last accessed January 2017)

[17] H. Hayasaki, I. Saitoh, K. Nakakura-Ohshima, et al, Tooth brushing for oral prophylaxis, Japan. Dent. Sci. Rev. 50 (2014) 69-77.h ttps://doi.org/10.1016/j.jdsr.2014.04.001

[18] G. Quigley, J. Hein, Comparative cleansing efficacy of manual and power brushing, J. Am. Dent. Assoc. 65 (1962) 26-29.h ttps://doi.org/10.14219/jada.archive.1962.0184

[19] S. Turesky, ND. Gilmore, I. Glickman, Reduced plaque formation by the chloromethyl analogue of vitamin C, J. Periodont. 41 (1970) 41-43h ttps://doi.org/10.1902/jop.1970.41.1.41

[20] Y. H. An, R. J. Friedman, Handbook of Bacterial Adhesion, 1st ed., Humana, Totowa, NJ, USA, 2000.h ttps://doi.org/10.1385/1592592244

[21] W. M. Dunne Jr., Bacterial adhesion: seen any good biofilms lately?, Clin. Microbiol. Rev. 15 (2002) 155-166.h ttps://doi.org/10.1128/CMR.15.2.155-166.2002

[22] H.-C. Flemming, J. Wingender, U. Szewzyk, Biofilm Highlights, 1st ed., J. W. Costerton (Ed.), Springer Series on Biofilms, Springer-Verlag, Berlin, 2011.h ttps://doi.org/10.1007/978-3-642-19940-0

[23] J. Schmidt, Ch. Zaugg, R. Weiger, C. Walter, Brushing without brushing? A review of the efficacy of powdered toothbrushes in noncontact biofilm removal, Clin. Oral Investig. 17 (2013) 687-709.h ttps://doi.org/10.1007/s00784-012-0836-8

[24] E. Bar-Zeev, K. R. Zodrow, S. E. Kwan, M. Elimelech, The importance of microscopic characterization of membrane biofilms in an unconfined environment, Desalination 348 (2014) 8-15.h ttps://doi.org/10.1016/j.desal.2014.06.003

[25]   B. H. Liu, K.-L. Li, K.-L. Kang, W.-K. Huang, J.-D. Liao, In situ biosensing of the nanomechanical property and electrochemical spectroscopy of Streptococcus mutans-containing biofilms, J. Phys. D: Appl. Phys. 46 (2013) 275401.h ttps://doi.org/10.1088/0022-3727/46/27/275401

[26]   http://fiji.sc/Auto_Threshold (last accessed January 2017)

[27]   http://www.icms.qmul.ac.uk/imaging/Protocols/Image%20analysis/Thresholding-.pdf (last accessed January 2017)

[28]   J. Schindelin, I. Arganda-Carreras, E. Frise, V. Kaynig, et al, Fiji: an open-source platform for biological-image analysis, Nature Methods 9 (2012) 676-682.h ttps://doi.org/10.1038/nmeth.2019

[29]   W.S.Rasband, ImageJ, U. S. National Institutes of Health, Bethesda, Maryland, USA, http://imagej.nih.gov/ij/, 1997 – 2012 (last accessed January 2017)

[30]   MA. Eid, YF. Talic, A clinical trial on the effectiveness of professional toothbrushing using dentifrice and water, Odontostomatol. Trop. 14 (1991) 9-12.

[31]   SP. Parizotto, CR. Rodrigues, J. da M. Singer, HC. Sef, Effectiveness of low cost toothbrushes, with or without dentifrice, in the removal of bacterial plaque in deciduous teeth, Pesqui. Odontol. Bras. 17 (2003) 17-23h ttps://doi.org/10.1590/S1517-74912003000100004

[32]   S. Paraskevas, PA. Versteeg, MF. Timmerman, U. van der Velden, GA. van der Weijden, The additional effect of dentifrices on the instant efficacy of toothbrushing, J. Periodontol. 77 (2006) 1522-1527.h ttps://doi.org/10.1902/jop.2006.050188

[33]   A. Binney, M. Addy, G. Newcombe, The plaque removal effects of single rinsing and brushings, J. Periodontol. 64 (1993) 181-185.h ttps://doi.org/10.1902/jop.1993.64.3.181

[34]   S. Paraskevas, N. Rosema, P. Versteeg, M. Timmerman, U. van der Velden, G. van der Weijden, The additional effect of a dentifrice on the instant efficacy of toothbrushing: a crossover study, J. Periodontol. 78 (2007) 1011–1016.h ttps://doi.org/10.1902/jop.2007.060339

[35]   G. Forward, Role of toothpastes in the cleaning of teeth Int. Dent. J. 41 (1991) 164–170.

[36]   J. C. Green, Oral hygiene and periodontal disease, Am. J. Public Health 53 (1963) 913-922.h ttps://doi.org/10.2105/AJPH.53.6.913

[37]   A. Haffajee, C. Smith, G. Torresyap, M. Thompson, D. Guerrero, S. Socransky, Efficacy of manual and powered toothbrushes. Effect on microbiological parameters, J. Clin. Periodontol. 28 (2001) 947-954.h ttps://doi.org/10.1034/j.1600-051x.2001.028010947.x

[38]   Delivering Better Oral Health: An evidence-based toolkit for prevention, Public Health England, 2007, https://www.gov.uk/government/uploads/system/uploads/attachment_data/file/367 563/DBOHv32014OCTMainDocument_3.pdf (last accessed January 2017)

[39]   T. Outhouse, R. Al-Alawi, Z. Fedorowicz, J. Keenan, Cochrane Database of Systematic Reviews (2006) Art. No.: CD005519

CHAPTER 8

Concluding Remarks

Since the first biomaterials and composites have been created in the labs worldwide, based on methods and systems existing in nature, researchers have been working on improving their approaches to better simulate the fascinating high-performance that natural materials exhibit. This book revealed my contribution to the research work in this field and it answered the question if biocompatible and biological materials can be influenced by external factors. The answers to this question are summarized here.

The nucleation and growth of biomimetically grown HA have been stimulated by the application of external factor in the face of high intensity pulsed laser beam. The approach mimics natural material formation processes and can be used to create structures organized on both micro- and nanometer-scale typical of what nature has developed, based on laser irradiation. The advantage is that the formation of the micrometer-scale architecture on the materials surface and the simultaneous nucleation of a CaP phase (HA precursor) occur within the time of the laser irradiation (< 5 min).

On the other side, organic substances were another factor used to influence the *in vitro* formation of HA and organosilicone polymer with biocompatible properties:

Pre-coating of materials with organic matrix based on non-collagenous ECM proteins influenced the HA morphology and long-term order of the mineral. The proteins guided the mineral nucleation, lowered the activation energy for the crystal nucleation and provided nucleation sites and oriented template for the mineral deposition as observed *in vivo*, thus leading to modulated physical properties of the grown HA.

The addition of diamond nanoparticles to the HA coating of metals improved the coating ductility and Vickers hardness without introducing residual stress and cracks. The composite coating was bioactive since it induced the formation of biomimetic HA with improved crystallinity. Adding ND particles to smooth plasma-polymerized hexamethyl-disiloxane layer slightly increased its surface roughness. Subsequent ammonia plasma treatment reduced its hydrophobicity thereby improving its interaction with living cells.

The formation of bacterial films on teeth and soft tissues in the mouth has been studied through the development of *in vitro* bacterial models. It was found that the physical effect of cavitation occurring around dental scaler instruments could be used to disrupt biofilms without actual contact, thus avoiding the mechanical contact with the dental tip which yields discomfort to the patient. Commercial toothpaste with granules and dual-face toothbrush with angled head were found sufficiently efficient in plaque removal from teeth and oral mucosa. These results may contribute to the prevention of dental diseases through improved oral health and patient comfort.

Keywords

About the Author

Dr. Emilia Pecheva

Dr. Emilia Pecheva is an Associated Professor and a Doctor of Sciences at the group of Biocompatible Materials, Institute of Solid State Physics, Bulgarian Academy of Sciences, Sofia, Bulgaria. She received her MSc degree in Laser physics from Sofia University and her PhD degree in solid state physics and materials science from the Bulgarian Academy of Sciences.

Prof. Pecheva has held several positions abroad: at the School of Dentistry, University of Birmingham, UK, the Institute of Biomaterials and Bioengineering, Tokyo Medical and Dental University, Japan, and the Institut d'Electronique du Solide et des Systemes, CNRS (now ICube, University of Strasbourg), Strasbourg, France. She has been working briefly with the industry on three grants with Unilever, Italy and Dentsply, USA.

Her fields of scientific research include biomineralization, biomaterials, hydroxyapatite, composite materials, dental hard tissue, surface modification, surface and interface characterization, laser interaction with materials, interaction material-aqueous media, polymers and cell-material interactions. Currently, Prof. Pecheva is the leader of the Biocompatible Materials group at the Institute of Solid State Physics, Bulgarian Academy of Sciences.